Commercial Oral Exam Guide

Fourth Edition

The comprehensive
guide to prepare you
for the FAA Oral Exam

by Michael D. Hayes

Aviation Supplies & Academics, Inc.
Newcastle, Washington

Commercial Oral Exam Guide
Fourth Edition

Aviation Supplies & Academics, Inc.
7005 132nd Place SE
Newcastle, Washington 98059-3153

© 1992–2000 Aviation Supplies & Academics, Inc.
All rights reserved. Published 1998. Third printing 2000.

None of the material in this guide supersedes any documents, procedures, or regulations issued by the Federal
Aviation Administration.

Printed in the United States of America

02   01   00   9   8   7   6   5   4   3

ISBN 1-56027-420-4
ASA-OEG-C4

Library of Congress Cataloging-in-Publication Data

Hayes, Michael D.
        Commercial oral exam guide / by Michael D. Hayes.
              p.   cm.
        "ASA-OEG-C"—T.p. verso.
        ISBN 1-56027-157-4
        1. Aeronautics—Examinations, questions, etc. 2. Aeronautics,
Commercial—Examinations, questions, etc. 3. United States.
Federal Aviation Administration—Examinations—Study guides.
4. Oral examinations. I. Title.
        TL546.5.H35   1992
        629.132'5217'076—dc20   92-36217
                                                  CIP

This guide is dedicated to the many talented students, pilots, and flight instructors I have had the opportunity to work with over the years. Also, special thanks to Mark Hayes and many others who supplied the patience, encouragement, and understanding necessary to complete the project.

— M.D.H.

Contents

Continued

Introduction

The *Commercial Oral Exam Guide* is a comprehensive guide designed for pilots who are involved in training for the Commercial Pilot Certificate. It was originally designed for use in a Part 141 flight school but quickly became popular with those training under 14 CFR Part 61 who are not affiliated with an approved school. The guide will also prove beneficial to pilots who wish to refresh their knowledge or who are preparing for a biennial flight review.

The Commercial Pilot Practical Test Standards book (FAA-S-8081-12A) specifies the areas in which knowledge must be demonstrated by the applicant before issuance of a pilot certificate or rating. The *Commercial Oral Exam Guide* is designed to evaluate a pilot's knowledge of those areas.

Commercial pilots are professionals engaged in various flight activities for compensation or hire. Because of their professional status, they should exhibit a significantly higher level of knowledge than a private pilot. This guide assumes that the pilot has the prerequisite knowledge necessary for Private Pilot Certification and attempts to cover only those advanced areas of knowledge necessary for Commercial Pilot Certification.

In this guide, questions and answers are organized into nine chapters which represent those areas of knowledge required for the practical test. At any time during the practical test, an examiner may ask questions pertaining to any of the subject areas within these divisions. Through intensive post-commercial-checkride debriefings, we have provided you with the most consistent questions asked, along with the information necessary for a knowledgeable response.

The guide may be supplemented with other comprehensive study materials as noted in parentheses after each question. For example: (AC 61-23C). The abbreviations for these materials and their titles are listed on the next page. Be sure that you use the latest revision of these references when reviewing for the test.

Continued

14 CFR Part 23	*Airworthiness Standards: Normal, Utility, Acrobatic, and Commuter Category Airplanes*
14 CFR Part 43	*Maintenance, Preventive Maintenance, Rebuilding, and Alteration*
14 CFR Part 61	*Certification: Pilots, Flight Instructors, and Ground Instructors*
14 CFR Part 91	*General Operating and Flight Rules*
NTSB Part 830	*Notification and Reporting of Aircraft Accidents and Incidents*
FAA-H-8083-1	*Aircraft Weight and Balance Handbook*
FAA-H-8083-3	*Airplane Flying Handbook*
FAA-P-8740-13	*Engine Operation for Pilots*
AC 00-6	*Aviation Weather*
AC 00-45	*Aviation Weather Services*
AC 20-32	*Carbon Monoxide (CO) Contamination in Aircraft*
AC 25-14	*High Lift and Drag Devices*
AC 61-23	*Pilot's Handbook of Aeronautical Knowledge*
AC 61-27	*Instrument Flying Handbook*
AC 61-65	*Certification: Pilots and Flight Instructors*
AC 65-12	*Airframe & Powerplant Mechanics: Powerplant Handbook*
AC 65-15	*Airframe & Powerplant Mechanics: Airframe Handbook*
AC 67-2	*Medical Handbook for Pilots*
AC 91-6	*Water, Slush and Snow on the Runway*
AC 91-33	*Use of Alternate Grades of Aviation Gasoline*
AC 91-67	*Minimum Equipment Requirements for General Aviation Operations under 14 CFR Part 91*
AC 120-12	*Private Carriage vs. Common Carriage of Persons or Property*
AIM	*Aeronautical Information Manual*
AFD	*Airport Facility Directory*
NOTAMs	*Notices to Airmen*
POH	*Pertinent Pilot Operating Handbooks*
AFM	*Airplane Flight Manuals*

Note: Be sure that you use the latest revision of these references when reviewing for the checkride.

A review of the information presented within this guide along with a general review of the *Private Oral Exam Guide* (ASA-OEG-P) should provide the necessary preparation for the oral section of an FAA Commercial Pilot checkride or recertification check.

Certificates
and Documents   **1**

A. Privileges and Limitations

1. What privileges apply to a commercial pilot? (14 CFR 61.133)

A person who holds a commercial pilot certificate may act as pilot-in-command of an aircraft:

a. Carrying persons or property for compensation or hire

b. For compensation or hire

Note: 14 CFR §61.133 also states that a commercial pilot must be qualified and comply with the applicable parts of the regulations that apply to the particular operation being conducted, for example Part 91 or 135.

2. Discuss commercial pilot operations.

A commercial pilot intending to conduct operations as a pilot-in-command of an aircraft carrying persons or property for compensation or hire should look cautiously at any proposal for revenue operating flights.

The following facts should be considered:

a. Part 61 states that you may be paid for acting as PIC of an aircraft engaged in carrying persons or property for compensation or hire. Part 61 does not mention, that if acting totally by yourself, you could be considered a commercial operator, and as such, be subject to an entirely different set of regulations.

b. A commercial pilot certificate by itself does not allow you to act as a commercial operator. It only allows you to work for a commercial operator and be paid for your service, with certain exceptions.

c. As a commercial pilot certain commercial operations are allowed without being in possession of an "Operating Certificate." Examples of such operations are: student instruction, certain nonstop sightseeing flights, ferry or training flights, aerial work operations including crop dusting, banner towing, aerial photography, powerline or pipeline patrol, etc. These operations are listed in 14 CFR §119.1.

3. What is meant by the term "commercial operator"? (14 CFR Part 1)

Commercial operator means a person who, for compensation or hire, engages in the carriage by aircraft in air commerce of persons or property, other than as an air carrier or foreign air carrier or under the authority of Part 375 of this title [Title 14]. Where it is doubtful that an operation is for "compensation or hire," the test applied is whether the carriage by air is merely incidental to the person's other business or is, in itself, a major enterprise for profit.

4. Define the term "common carriage." (AC 120-12A)

Common carriage refers to the carriage of passengers or cargo as a result of advertising the availability of the carriage to the public. A carrier becomes a common carrier when it "holds itself out" to the public, or a segment of the public, as willing to furnish transportation within the limits of its facilities to any person who wants it. There are four elements in defining a "common carrier":

a. A holding out or a willingness to

b. Transport persons or property

c. From place to place

d. For compensation.

5. Define the term "holding out." (AC 120-12A)

Holding out implies offering to the public the carriage of persons and property for hire either intrastate or interstate. This holding out which makes a person a common carrier can be done in many ways, and it does not matter how it is done.

a. Signs and advertising are the most direct means of holding out but are not the only ones.

b. A holding out may be accomplished through the actions of agents, agencies, or salesmen who may obtain passenger traffic from the general public and collect them into groups to be carried by the operator.

c. Physically holding out without advertising, yet gaining a reputation to "serve all," is sufficient to constitute an offer to carry all customers. Physical holding out may take place in many ways. For example, the expression of willingness to all customers with whom contact is made that the operator can and will perform the requested service is sufficient. It doesn't matter if the holding out generates little success. The important issue is the nature and character of the operation.

d. A carrier holding itself out as generally willing to carry only certain kinds of traffic is nevertheless a common carrier.

6. Define the term "private carriage." (AC 120-12A)

Carriage for hire which does not involve holding out is "private carriage." Private carriage for hire is carriage for one or several selected customers, generally on a long-term basis. The number of contracts must not be too great, otherwise it implies a willingness to make a contract with anybody. A carrier operating with 18 to 24 contracts has been labeled a common carrier because it has held itself out to serve the general public to the extent of its facilities. Private carriage has been found in cases where three contracts have been the sole basis of the operator's business.

Note: It should be understood that the number of contracts is not the determining factor when assessing whether a particular operation is common carriage or private carriage. Persons intending to conduct only private operations in support of other business should look cautiously at any proposal for revenue-generating flights which would most likely require certification as an air carrier.

7. State some typical examples of private carriage operations. (AC 120-12A)

a. Carriage of property for hire for one customer in intrastate commerce

b. Carriage of persons or property for hire for a few selected customers on an intrastate basis

c. Carriage of cargo for an industrial firm on an intrastate basis

8. **Determine if either of the following two scenarios are common carriage operations and, if so, why?**

   Scenario 1: **I am a local businessman and require a package to be flown to a distant destination ASAP. I will pay you to fly my airplane to deliver this package.**

   Scenario 2: **I am a local businessman and require a package to be flown to a distant destination ASAP. You reply that you can do the job for a fee. You promptly line up a local rental aircraft you're checked out in and deliver the package.**

   Scenario 2 would be considered a common carriage operation because you are holding out by indicating a general willingness to all customers with whom contact is made to transport persons or property from place to place for compensation.

9. **Briefly describe the Federal Aviation Regulations Parts 119, 121, 125, 135, and 137.**

   Part 119—Certification: Air Carriers and Commercial Operators
   Part 121—Operating Requirements: Domestic, Flag, and Supplemental Operations
   Part 125—Certification and Operations: Airplanes having a seating capacity of 20 or more passengers or a maximum payload capacity of 6,000 pounds or more
   Part 135—Operating Requirements: Commuter and On-Demand Operations
   Part 137—Agricultural Aircraft Operations

10. **What limitation is imposed on a newly certificated commercial airplane pilot if that person does not hold an instrument rating?** (14 CFR 61.133)

    The pilot must hold an instrument rating in the same category and class, or the Commercial Pilot Certificate that is issued is endorsed with a limitation prohibiting the following:

    a. The carriage of passengers for hire in airplanes on cross-country flights in excess of 50 nautical miles;

    b. The carriage of passengers for hire in airplanes at night.

11. To act as pilot-in-command or in any other capacity as a required flight crewmember of a civil aircraft, what must a pilot have in his/her physical possession or readily accessible in the aircraft? (14 CFR 61.3)

    a. A valid pilot certificate
    b. A current and appropriate medical certificate

12. If a certificated pilot changes his/her permanent mailing address and fails to notify the FAA Airman Certification branch of the new address, the pilot is entitled to exercise the privileges of the pilot certificate for what period of time? (14 CFR 61.60)

30 days after the date of the move.

13. If a pilot certificate were accidentally lost or destroyed, a pilot could continue to exercise privileges of that certificate provided he/she follows what specific procedure? (14 CFR 61.29)

    a. An application for the replacement of a lost or destroyed airman certificate issued under Part 61 is made by letter to the Department of Transportation, Federal Aviation Administration; and

    b. A person who has lost a certificate may obtain a facsimile from the FAA confirming that it was issued. The facsimile may be carried as a certificate for up to 60 days pending receipt of a duplicate certificate.

14. To act as pilot-in-command of a high-performance aircraft, what flight experience requirements must be met? (14 CFR 61.31)

A high-performance airplane is an airplane with an engine of more than 200 horsepower. To act as pilot-in-command of a high-performance airplane a person must have:

    a. Received and logged ground and flight training from an authorized instructor in a high-performance airplane, or in a flight simulator or flight training device that is representative of a high-performance airplane; and

Continued

b. Been found proficient in the operation and systems of the airplane; and

c. Received a one-time endorsement in the pilot's logbook from an authorized instructor who certifies the person is proficient to operate a high-performance airplane.

15. To act as pilot-in-command of a pressurized aircraft, what flight experience requirements must be met? (14 CFR 61.31)

To act as pilot-in-command of a pressurized aircraft (an aircraft that has a service ceiling or maximum operating altitude, whichever is lower, above 25,000 feet MSL), a person must have received and logged ground and flight training from an authorized instructor and obtained an endorsement in the person's logbook or training record from an authorized instructor who certifies the person has:

a. satisfactorily accomplished the ground training which includes high-altitude aerodynamics, meteorology, respiration, hypoxia, etc.; and

b. received and logged training in a pressurized aircraft, or in a flight simulator or flight training device representative of a pressurized aircraft, and obtained an endorsement in the person's logbook or training record from an authorized instructor who found the person proficient in the operation of pressurized aircraft (must include normal cruise flight above 25,000 feet MSL, emergency procedures for rapid decompression, emergency descent procedures).

16. To act as pilot-in-command of a tailwheel airplane, what flight experience requirements must be met? (14 CFR 61.31)

No person may act as pilot-in-command of a tailwheel airplane unless that person has received and logged flight training from an authorized instructor in a tailwheel airplane and received an endorsement in the person's logbook from an authorized instructor who found the person proficient in the operation of a tailwheel airplane. The flight training must include at least the following maneuvers and procedures: normal and crosswind takeoffs and landings, wheel landings and go-around procedures.

17. When would a commercial pilot be required to hold a type rating? (14 CFR 61.31)

According to 14 CFR §61.31, a person who acts as a pilot-in-command of any of the following aircraft, must hold a type rating for that aircraft:

a. Large aircraft (gross weight over 12,500 pounds, except lighter-than-air)

b. Turbojet-powered airplanes

c. Other aircraft specified by the Administrator through aircraft type certificate procedures.

18. With respect to certification, privileges, and limitations of airmen, define the terms "Category," "Class" and "Type." (14 CFR Part 1)

Category—a broad classification of aircraft; i.e., airplane, rotor-craft, glider, etc.

Class—a classification of aircraft within a category having similar operating characteristics; i.e., single-engine land, multi-engine land, etc.

Type—a specific make and basic model of aircraft including modifications that do not change its handling or flight characteristics; i.e., DC-9, B-737, etc.

19. Can a pilot with a commercial certificate and multi-engine land rating carry passengers in a single-engine airplane? (14 CFR 61.31)

No. Unless he holds a category and class rating for that aircraft, a person may not act as pilot-in-command of an aircraft that is carrying another person or is operated for compensation or hire.

20. Can a commercial pilot carry a passenger in an aircraft operated in formation flight? (14 CFR 91.111)

No person may operate an aircraft, carrying passengers for hire, in formation flight.

21. Can a commercial pilot carry passengers in a restricted, limited or experimental category aircraft?
(14 CFR 91.313, 91.315, 91.317, and 91.319)

No person may operate a restricted, limited, or experimental category aircraft carrying persons or property for hire.

22. When may a commercial pilot log flight time as second-in-command time? (14 CFR 61.51)

According to 14 CFR §61.51, a pilot may log second-in-command time only for that flight time during which that person:

1. Is qualified according to the second-in-command requirements of 14 CFR §61.55, and occupies a crewmember station in an aircraft that requires more than one pilot by the aircraft's type certificate; or

2. Holds the appropriate category, class, and instrument rating (if an instrument rating is required for the flight) for the aircraft being flown, and more than one pilot is required under the type certification of the aircraft or the regulations under which the flight is being conducted.

23. You are currently en route to your destination and the sun has set. When can you begin logging flight time as "night" flight time? (14 CFR Part 1)

"Night" is defined as the time between the end of evening civil twilight and the beginning of morning civil twilight, as published in the American Air Almanac and converted to local time. All flight time that occurs during this period of time is considered "night" flight time.

B. Currency Requirements

1. What are the requirements to remain current as a commercial pilot? (14 CFR 61.56 and 61.57)

a. To remain current a commercial pilot must have accomplished a flight review given in an aircraft for which that pilot is rated by an appropriately-rated instructor within the preceding 24 calendar months.

b. To carry passengers, a pilot must have made within the preceding 90 days:

   i. Three takeoffs and three landings as the sole manipulator of the flight controls of an aircraft of the same category and class and, if a type rating is required, of the same type.

   ii. If the aircraft is a tailwheel airplane, the landings must have been made to a full stop.

   iii. If operations are to be conducted during the period beginning 1 hour after sunset or 1 hour before sunrise, with passengers on board, the pilot-in-command must have made at least three takeoffs and three landings to a full stop during that period in an aircraft of the same category, class, and type (if a type rating is required).

Note: A person may act as pilot-in-command of a flight under day VFR or day IFR if no persons or property are carried if the flight review is current.

2. What class of medical certificate is required for commercial pilots? (14 CFR 61.23)

A second-class medical certificate is required in order to exercise commercial pilot privileges.

3. What is the duration of a second-class medical certificate for operations requiring a commercial pilot certificate? (14 CFR 61.23)

A second-class medical certificate expires at the end of the last day of the 12th month after the month of the date of the examination shown on the certificate for operations requiring a commercial certificate.

4. Is a commercial pilot required to log all flight time?
(14 CFR 61.51)

Each person must document and record, in a manner acceptable to the Administrator, the training and aeronautical experience used to meet the requirements for a certificate, rating or flight review of this part. They must also document and record the aeronautical experience required for meeting the recent flight experience requirements of this part.

C. Aircraft Certificates and Documents

1. What documents are required on board an aircraft prior to flight? (14 CFR 91.9 and 91.203)

A irworthiness Certificate

R egistration Certificate

O wner's manual or operating limitations

W eight and balance data

2. Which documents, required on board an aircraft, must be displayed in such a way so as to be visible by both passengers and crew? (14 CFR 91.203)

No person may operate a civil aircraft unless the Airworthiness Certificate required or a special flight authorization issued is displayed at the cabin entrance or cockpit entrance so that it is legible to passengers and crew.

3. Are the aircraft and engine logbooks required to be carried on board an aircraft?

No. Generally, it is more advisable to keep the logbooks in a safe, secure place such as the office, home, etc. The regulations do not specifically state where the logbooks are to be kept, but they do say that they should be made available upon request.

4. How can a pilot determine if his/her aircraft is equipped with a Mode C altitude encoding transponder?

By referencing the current weight and balance equipment list for that aircraft, a pilot could positively determine if a Mode C transponder is installed.

5. If the Airworthiness Certificate of a particular aircraft indicated one of the following categories, what significance would this have? (14 CFR Part 23)
   a. Normal Category
   b. Utility Category

   a. *Normal category*—Aircraft structure capable of withstanding a load factor of 3.8 Gs without structural failure. Applicable to aircraft intended for non-aerobatic operation.

   b. *Utility category*—Aircraft structure must be capable of withstanding a load factor of 4.4 Gs. This would usually permit limited aerobatics, including spins (if approved for the aircraft).

6. How can a pilot determine if the approved airplane flight manual (AFM) is on board the aircraft? (AC 60-6B)

The aircraft manufacturer's approved AFM will always contain a title page with the specific aircraft registration and serial number. If this information is not included, the manual is a reproduction (called a "Pilot's Operating Handbook"), and can be used for general study purposes only. Pilots can check the airplane's Type Certificate Data Sheet to determine if an AFM exists for a particular airplane. If an AFM is not available for the aircraft, 14 CFR §91.9 says that manual material, markings, placards, or a combination thereof must be on board.

7. What are "Special Flight Permits," and when are they necessary? (14 CFR 91.213 and 21.197)

A "Special Flight Permit" may be issued for an aircraft that may not currently meet applicable airworthiness requirements but is capable of safe flight. These permits are typically issued for the following purposes:

a. Flying an aircraft to a base where repairs, alterations or maintenance are to be performed or to a point of storage.

b. Delivering or exporting an aircraft.

c. Production flight testing new production aircraft.

d. Evacuating aircraft from areas of impending danger.

e. Conducting customer demonstration flights in new production aircraft that have satisfactorily completed production flight tests.

D. Aircraft Maintenance Requirements

1. Who is responsible for ensuring that an aircraft is maintained in an airworthy condition? (14 CFR 91.403)

The owner or operator of an aircraft is primarily responsible for maintaining an aircraft in an airworthy condition.

2. After aircraft inspections have been made and defects have been repaired, who is responsible for determining that the aircraft is in an airworthy condition? (14 CFR 91.7)

The pilot-in-command of a civil aircraft is responsible for determining whether that aircraft is in condition for safe flight. The pilot-in-command shall discontinue the flight when unairworthy mechanical, electrical, or structural conditions occur.

3. Can you legally fly an aircraft that has an inoperative flap position indicator?

Unless operations are conducted under 14 CFR §91.213, the regulations require that all equipment installed on an aircraft in compliance with either the Airworthiness Standards or the Operating Rules must be operative. If equipment originally installed in the aircraft is no longer operative, the Airworthiness Certificate is not valid until such equipment is either repaired or removed from that aircraft. However, the rules also permit the publication of a Minimum Equipment List (MEL) where compliance with these equipment requirements is not necessary in the interest of safety under all conditions.

4. Can an aircraft operator allow flight operations to be conducted in an aircraft with known inoperative equipment? (AC 91-67, 14 CFR 91.213)

Part 91 describes acceptable methods for the operation of an aircraft with certain inoperative instruments and equipment which are not essential for safe flight. These acceptable methods of operation are:

1. Operation of aircraft with a Minimum Equipment List (MEL), as authorized by 14 CFR §91.213(a).

2. Operation of aircraft without a MEL under 14 CFR §91.213(d).

5. What are "Minimum Equipment Lists"? (AC 91-67)

The Minimum Equipment List (MEL) is a precise listing of instruments, equipment, and procedures that allows an aircraft to be operated under specific conditions with inoperative equipment. The MEL is the specific inoperative equipment document for a particular make and model aircraft by serial and registration numbers; e.g., BE-200, N12345. The FAA-approved MEL includes only those items of equipment which the administrator finds may be inoperative and yet maintain an acceptable level of safety by appropriate conditions and limitations.

6. If an aircraft is not being operated under a MEL, how can you determine which instruments and equipment on board can be inoperative and the aircraft still be legal for flight? (14 CFR 91.213)

A person may takeoff an aircraft in operations conducted under Part 91 with inoperative instruments and equipment without an approved Minimum Equipment List provided the inoperative instruments and equipment are not —

Continued

a. Part of the VFR-day type certification instruments and equipment prescribed in the applicable airworthiness regulations under which the aircraft was type certificated;

b. Indicated as required on the aircraft's equipment list, or on the Kinds of Operations Equipment List, for the kind of flight operation being conducted.

7. What is an aircraft equipment list, and where is it found? (AC 91-67)

The aircraft equipment list is an inventory of equipment installed by the manufacturer or operator on a particular aircraft. It is usually found with the weight and balance data.

8. What length of time can an aircraft be flown with inoperative equipment on board? (AC 91-67)

An operator may defer maintenance on inoperative equipment that has been deactivated or removed and placarded inoperative. When the aircraft is due for inspection in accordance with the regulation, the operator should have all inoperative items repaired or replaced. If an owner does not want specific inoperative equipment repaired, then the maintenance person must check each item to see if it conforms to the requirements of 14 CFR §91.213. The maintenance person must ensure that each item of inoperative equipment that is to remain inoperative is placarded appropriately.

9. What regulations apply concerning the operation of an aircraft that has had alterations or repairs which may have substantially affected its operation in flight? (14 CFR 91.407)

No person may operate or carry passengers in any aircraft that has undergone maintenance, preventive maintenance, rebuilding, or alteration that may have appreciably changed its flight characteristics or substantially affected its operation in flight until an appropriately-rated pilot with at least a private pilot certificate

a. flies the aircraft,

b. makes an operational check of the maintenance performed or alteration made, and

c. logs the flight in the aircraft records.

10. How long does the Airworthiness Certificate of an aircraft remain valid? (14 CFR Part 21)

Standard Airworthiness Certificates are effective as long as the maintenance, preventive maintenance, and alterations are performed in accordance with Parts 43 and 91 and the aircraft is registered in the United States.

11. What are the required maintenance inspections for aircraft? (14 CFR 91.409)

a. Annual inspection—within the preceding 12 calendar months

b. 100-hour inspection—if carrying any person (other than a crewmember) for hire or giving flight instruction for hire.

Note: If an aircraft is operated for hire, it must have a 100-hour inspection as well as an annual inspection when due. If not operated for hire, it must have an annual inspection only.

12. Can a 100-hour inspection be substituted for an annual inspection? (14 CFR 91.409)

No, an annual inspection is acceptable as a 100-hour inspection, but the reverse is not true. The 100-hour inspection is generally the same as an annual inspection but is not considered as intense an inspection as the annual. Also, the 100-hour inspection can be signed off by an Airframe & Powerplant Mechanic (A&P), but the annual inspection must be signed off by an Aircraft Inspector (IA).

13. What types of aircraft inspections can be substituted for a 100-hour inspection? (14 CFR 91.409)

The following may replace a 100-hour inspection:

a. Aircraft inspected in accordance with an approved aircraft inspection program under Part 125, 127, or 135.

b. Progressive inspections which provide for the complete inspection of an aircraft by specifying the intervals in hours and days when routine and detailed inspections will be performed during a 12-calendar month period.

c. Inspection programs approved by the Administrator for large aircraft, turbojet multi-engine airplanes, turboprop-powered multi-engine airplanes and turbine powered rotorcraft.

14. If an aircraft, carrying passengers for hire, has been on a schedule of inspection every 100 hours, under what condition may it continue to operate beyond the 100 hours without a new inspection? (14 CFR 91.409)

The 100-hour limitation may be exceeded by not more than 10 hours while en route to reach a place where the inspection can be done. The excess time used to reach a place where the inspection can be done must be included in computing the next 100 hours of time in service.

15. What are the required tests and inspections of aircraft and equipment to be legal for both VFR and IFR flights? (14 CFR 91.171, 91.203, 91.411, and 91.413)

a. The aircraft must have an annual inspection. If operated for hire or rental, it must also have a 100-hour inspection. A record must be kept in the aircraft/engine logbooks.

b. The pitot/static system must be checked within the preceding 24 calendar months. A record must be kept in the aircraft logbook.

c. The transponder must have been checked within the preceding 24 calendar months. A record must be kept in the aircraft logbook.

d. The altimeter must have been checked within the preceding 24 calendar months. A record must be kept in the aircraft logbook.

e. The VOR must have been checked within the preceding 30 days. A record must be kept in a bound logbook.

f. If operations require an emergency locator transmitter (ELT), it must be inspected within 12 calendar months after the last inspection.

Note: Be capable of locating the last 100-hour/annual inspections in the aircraft and engine logbooks and be able to determine when the next inspections are due. Also, be capable of locating all required inspections for instruments and equipment necessary for legal VFR/IFR flight.

16. Define "preventive maintenance." (14 CFR Part 43)

"Preventive maintenance" means simple or minor preservation operations and the replacement of small standard parts not involving complex assembly operations. Certificated pilots, excluding student pilots, may perform preventive maintenance on any aircraft owned or operated by them (excluding aircraft used in air carrier service). Examples of such operations are: oil changes, wheel bearing lubrication and hydraulic fluid refills. A record of preventive maintenance must be entered in the appropriate records.

17. What are "Airworthiness Directives"?

An "AD" is the medium used by the FAA to notify aircraft owners and other potentially interested persons of unsafe conditions that may exist because of design defects, maintenance, or other causes, and to specify the conditions under which the product may continue to be operated. ADs are regulations, and compliance is mandatory. It is the aircraft owner's or operator's responsibility to ensure compliance with all pertinent ADs.

Additional Study Questions

1. As a newly certificated commercial pilot, you are ready to utilize your certificate. Can you begin charging for your services? (AC 120-12)

2. As a commercial pilot, being paid for your services, what Federal Aviation Regulations must you comply with?

3. To act as pilot-in-command of an aircraft during IFR operations under Part 135, what minimum experience is required?

4. Does an aircraft registration certificate have an expiration date? (14 CFR Part 43)

5. What length of time is a temporary registration certificate valid? (14 CFR Part 43)

6. How are "Special Flight Permits" obtained? (14 CFR §21.197)

7. How can a pilot determine if all Airworthiness Directives have been complied with for a particular airplane? (14 CFR Part 43)

8. Aircraft maintenance records must include what information? (14 CFR §91.417)

9. Are Minimum Equipment Lists available for all aircraft? (AC 91-67)

10. How can a pilot determine which placards are required to be displayed in a particular airplane? (POH/AFM)

Weather

A. Weather Briefing Services

1. What is the primary means of obtaining a weather briefing? (AIM 7-1-2)

The primary source of preflight weather briefings is an individual briefing obtained from a briefer at the AFSS/FSS. These briefings, which are tailored to your specific flight, are available 24 hours a day through the use of the toll free number (1-800-WX BRIEF).

2. What are some examples of other sources of weather information? (AIM 7-1-2)

a. Telephone Information Briefing Service (TIBS) (AFSS)

b. Transcribed Weather Broadcasts (TWEB)

c. Telephone Access to TWEB (TEL-TWEB)

d. Weather and aeronautical information from numerous private industry sources.

e. The Direct User Access System (DUATS)

3. Where can you find a listing of FSS and weather information numbers? (AIM 7-1-2)

Numbers for these services can be found in the Airport/Facility Directory under "FAA and NWS Telephone Numbers" section. They are also listed in the U.S. Government section of the local telephone directory.

4. What type of weather briefings are available from a FSS briefer? (AIM 7-1-3)

Standard Briefing—Request anytime you are planning a flight and you have not received a previous briefing or have not received preliminary information through mass-dissemination media; e.g., TIBS, TWEB, etc.

Abbreviated Briefing—Request when you need information to supplement mass-disseminated data, update a previous briefing, or when you need only one or two items.

Continued

Outlook Briefing—Request whenever your proposed time of departure is six or more hours from the time of the briefing. This is for planning purposes only.

Inflight Briefing—Request when needed to update a preflight briefing.

5. What pertinent information should a weather briefing include? (AIM 7-1-3)

a. Adverse Conditions

b. VFR Flight Not Recommended

c. Synopsis

d. Current Conditions

e. Enroute Forecast

f. Destination Forecast

g. Winds Aloft

h. Notices to Airmen

i. ATC Delay

j. Pilots may obtain the following from AFSS/FSS briefers upon request: Information on MTRs and MOAs, a review of printed NOTAM publication, approximate density altitude information, information on air traffic services and rules, customs/immigration procedures, ADIZ rules, search and rescue, LORAN-C NOTAMs, GPS RAIM availability, and other assistance as required.

6. What is "EFAS"? (AIM 7-1-4)

En route Flight Advisory Service (EFAS) is a service specifically designed to provide enroute aircraft with timely and meaningful weather advisories pertinent to the type of flight intended, route of flight, and altitude. In conjunction with this service, EFAS is also a central collection and distribution point for pilot reported weather information (PIREPs). EFAS provides communications capabilities for aircraft flying at 5,000 feet AGL to 17,500 feet MSL on a common frequency of 122.0 MHz. It is also known as "Flight Watch."

7. What is "HIWAS"? (AIM 7-1-9)

Hazardous In-flight Weather Advisory Service (HIWAS) is a continuous broadcast of in-flight weather advisories including summarized Aviation Weather Warnings, SIGMETs, Convective SIGMETs, Center Weather Advisories, AIRMETs, and urgent PIREPs. HIWAS is an additional source of hazardous weather information which makes this data available on a continuous basis.

B. Weather Reports and Forecasts

1. What is a "METAR"? (AC 00-45E)

METAR—Aviation Routine Weather Report; an hourly surface observation of conditions observed at an airport. A METAR report contains the following sequence of elements: Type of report (Routine or Special) station designator, time of report, wind, visibility, weather and obstructions to visibility, sky conditions, temperature/dew point, altimeter setting and remarks.

Example:
METAR KBNA 1250Z 33018KT 290V360 1/2SM R31/2700FT
+SN BLSNFG VV008 00/M03 A2991 RMK RAE42SNB42

2. What are PIREPs (UA), and where are they usually found? (AC 00-45E)

An abbreviation for "Pilot Reports," they contain information concerning weather as observed by pilots enroute. Required elements for all PIREPs are message type, location, time, flight level, type of aircraft, and at least one weather element encountered. All altitudes are MSL unless otherwise noted. Distances are in nautical miles, and time is in UTC. A PIREP is usually transmitted as an individual report but can be appended to a surface aviation weather report or placed into collectives.

3. What are "Radar Weather Reports" (SD)? (AC 00-45E)

Thunderstorms and general areas of precipitation can be observed by radar. Most radar stations report each hour at H+35 with intervening special reports as required. The report includes the type, intensity, intensity trend and location of the precipitation. Also included is the echo top of the precipitation and if significant, the base echo. All heights are reported above Mean Sea Level (MSL). Radar Weather Reports help pilots plan ahead to avoid thunderstorm areas. Once airborne, however, they must depend on visual sighting or airborne radar to evade individual storms.

4. What are Terminal Aerodrome Forecasts (TAFs)? (AC 00-45E)

A TAF is a concise statement of the expected meteorological conditions at an airport during a specified period (usually 24 hours). TAFs are issued in the following format: Type (routine or amended), location identifier, issuance time, valid time and the forecast with the basic format being WIND/VISIBILITY/ WEATHER/SKY CONDITION. Forecasts are issued three times daily for the next 24-hour period.

5. What is an "Aviation Area Forecast" (FA)? (AC 00-45E)

An Aviation Area Forecast (FA) is a forecast of general weather conditions over an area the size of several states. It is used to determine forecast en route weather and to interpolate conditions at airports which do not have TAFs issued. FAs are issued three times a day for each of the contiguous 48 states. Contains a 12-hour specific forecast, followed by a 6-hour (18-hour in Alaska) categorical outlook for a total forecast period of 18 hours (30 hours in Alaska).

6. What are In-Flight Aviation Weather Advisories (WST, WS, WA)? (AC 00-45E)

In-flight Aviation Weather Advisories are forecasts to advise en route aircraft of development of potentially hazardous weather. All heights are referenced to MSL, except in the case of ceilings (CIG) which indicates above ground level. The advisories are of three types: Convective SIGMET (WST), SIGMET (WS), and AIRMET (WA).

7. What is a convective SIGMET? (AC 00-45E)

Convective SIGMETs (WST) imply severe or greater turbulence, severe icing and low-level wind shear. They may be issued for any convective situation which the forecaster feels is hazardous to all categories of aircraft. Convective SIGMET bulletins are issued for the Eastern (E), Central (C) and Western (W) United States (Convective SIGMETs are not issued for Alaska or Hawaii). Bulletins are issued hourly at H+55. Special bulletins are issued at any time as required and updated at H+55. The text of the bulletin consists of either an observation and a forecast or just a forecast. The forecast is valid for up to 2 hours.

a. Severe thunderstorm due to:

    1. Surface winds greater than or equal to 50 knots.

    2. Hail at the surface greater than or equal to ¾ inches in diameter.

    3. Tornadoes.

b. Embedded thunderstorms.

c. A line of thunderstorms.

d. Thunderstorms greater than or equal to VIP level 4 affecting 40 percent or more of an area at least 3,000 square miles.

8. What is a SIGMET (WS)? (AC 00-45E)

A SIGMET (WS) advises of non-convective weather that is potentially hazardous to all aircraft. SIGMETs are issued for the six areas corresponding to the FA areas. The maximum forecast period is four hours. In the conterminous U.S., SIGMETs are issued when the following phenomena occur or are expected to occur:

a. Severe icing not associated with a thunderstorm.

b. Severe or extreme turbulence or clear air turbulence (CAT) not associated with thunderstorms.

c. Duststorms, sandstorms or volcanic ash lowering surface or in-flight visibilities to below 3 miles.

d. Volcanic eruption.

9. What is an AIRMET (WA)? (AC 00-45E)

AIRMETs (WA) are advisories of significant weather phenomena but describe conditions at intensities lower than those which trigger SIGMETs. AIRMETS are intended for dissemination to all pilots in the preflight and enroute phase of flight to enhance safety. AIRMET bulletins are issued on a scheduled basis every 6 hours and contain details on one or more of the following phenomena when they occur or are forecast to occur:

a. Moderate icing

b. Moderate turbulence

c. Sustained surface winds of 30 knots or more

d. Ceilings less than 1,000 feet and/or visibilities less than 3 miles affecting over 50 percent of the area at one time

e. Extensive mountain obscurement

10. What is a "Winds and Temperatures Aloft" forecast (FD)? (AC 00-45E)

Winds and temperatures aloft are forecast for specific locations in the contiguous U.S. and also for a network of locations in Alaska and Hawaii. Forecasts are made twice daily based on 00Z and 12Z data for use during specific time intervals. A 4-digit group will show wind direction, in reference to true north, and wind speed in knots. A 6-digit group will include forecast temperatures in degrees Celsius.

11. What valuable information can be determined from Winds and Temperatures Aloft forecasts (FD)? (AC 00-45E)

Most favorable altitude—based on winds and direction of flight.

Areas of possible icing—by noting air temperatures of +2°C to -20°C.

Temperature inversions.

Turbulence—by observing abrupt changes in wind direction and speed at different altitudes.

12. What are "Center Weather Advisories" (CWA)?
(AC 00-45E)

They are short-range forecasts, or *nowcasts*, for use by pilots and air traffic controllers, of possible adverse weather conditions anticipated in the center's area of coverage (en route and terminal areas). CWAs are unscheduled forecasts concerned with conditions anticipated within the next two hours of the time of the advisory, and are valid for a maximum of two hours, for that particular ARTCC.

13. What is a "Convective Outlook" (AC)? (AC 00-45E)

A Convective Outlook (AC) describes the prospects for general thunderstorm activity during the following 24 hours. Areas in which there is high, moderate, or slight risk of severe thunderstorms are included as well as areas where thunderstorms may approach severe limits. Outlooks are transmitted at 0700Z and 1500Z and are valid until 1200Z the next day.

C. NOTAMs

1. What are NOTAMs? (AIM 5-1-3)

Notices To Airmen (NOTAM)—Time critical aeronautical information which is of either a temporary nature or not sufficiently known in advance to permit publication on aeronautical charts or in other operational publications receives immediate dissemination via the National NOTAM System. It includes such information as airport or primary runway closures, changes in the status of navigational aids, ILS's, radar service availability, and other information essential to planned enroute, terminal, or landing operations.

2. What are the three categories of NOTAMs? (AIM 5-1-3)

There are three types of NOTAMs generated by the FAA:

a. NOTAM (D)—A NOTAM given (in addition to local dissemination) distant dissemination beyond the area of responsibility of the Flight Service Station. These NOTAMs will be stored

Continued

and available until canceled. NOTAM Ds contain information on all civil public use airports and navigational facilities that are part of the National Airspace System. NOTAM D items are serious enough to affect whether or not an airport or a certain facility is usable.

b. NOTAM(L)—A NOTAM given local dissemination by voice and other means to satisfy local user requirements. NOTAM(L) information may include conditions such as taxiway closures, persons and/or equipment near or crossing runways, airport rotating beacon outages and other information that would have little impact on non-local operations.

c. FDC NOTAM—The National Flight Data Center will issue these NOTAMS when it becomes necessary to disseminate information which is regulatory in nature. FDC NOTAMs contain such things as amendments to published IAPs and other current aeronautical charts. They are also used to advertise temporary flight restrictions caused by such things as natural disasters or large scale public events that may generate congestion of air traffic over a site.

3. What particular type of NOTAMs will be omitted in a pilot briefing if not specifically requested by the pilot? (AIM 7-1-3)

NOTAM (D) information and FDC NOTAMs which have been published in the Notices to Airmen Publication (NTAP) are not included in pilot briefings unless a review of this publication is specifically requested by the pilot. For complete flight information you are urged to review the printed NOTAMs in the Notices to Airmen publication and the Airport Facilities Directory in addition to obtaining a briefing. The NOTAM publication cycle is every 28 days.

4. Where can NOTAM information be obtained? (AIM 5-1-3)

a. Nearest FSS

b. Airport/Facility Directory

c. Locally broadcast ATIS

d. Hourly Surface observations

e. Notice to Airmen Publication (NTAP)—Printed NOTAMs; not normally provided in a briefing; must make specific request for.

D. Aviation Weather Hazards

1. What are the two major classifications of thunderstorms? (AC 00-6A)

Air mass thunderstorms—Most often result from surface heating. They occur at random in unstable air and last for only an hour or two. They reach maximum intensity and frequency over land during middle and late afternoon. Off-shore they reach a maximum during late hours of darkness when land temperature is coolest and cool air flows off the land over the relatively warm water.

Steady-state thunderstorms—Usually form in lines, last for several hours, dump heavy rain and possibly hail, and produce strong gusty winds and possibly tornadoes. They are normally associated with weather systems. Fronts, converging winds, and troughs aloft force upward motion spawning these storms which often form into squall lines. They are intensified by afternoon heating.

2. What procedures should be followed when avoiding turbulence around thunderstorms? (AIM 7-1-26)

Above all, remember this: Never regard any thunderstorm lightly even when radar observers report the echoes are of light intensity. Avoiding thunderstorms is the best policy. The following are some "Do's and Don'ts" of thunderstorm avoidance:

a. Don't land or take off in the face of an approaching thunderstorm. A sudden gust front of low-level turbulence could cause loss of control.

b. Don't attempt to fly under a thunderstorm even if you can see through to the other side. Turbulence and wind shear under the storm could be disastrous.

c. Don't fly without airborne radar into a cloud mass containing scattered embedded thunderstorms. Scattered thunderstorms not embedded can usually be visually circumnavigated.

d. Don't trust the visual appearance to be a reliable indicator of the turbulence inside a thunderstorm.

e. Do avoid, by at least 20 miles, any thunderstorms identified as severe or giving an intense radar echo. This is especially true under the anvil of a large cumulonimbus cloud.

Continued

f. Do clear the top of a known or suspected severe thunderstorm by at least 1,000 feet altitude for each 10 knots of wind speed at the cloud top. (*Note:* This should exceed the altitude capability of most aircraft.)

g. Do circumnavigate the entire area if the area has $^6/_{10}$ thunderstorm coverage.

h. Do remember that vivid and frequent lightning indicates the probability of a severe thunderstorm.

i. Do regard as extremely hazardous any thunderstorm with tops of 35,000 feet or higher whether the top is visually sighted or determined by radar.

3. Can ATC provide in-flight assistance in avoiding thunderstorms and severe weather? (AIM 7-1-12)

Yes, to the extent possible, controllers will issue pertinent information on weather or chaff areas and assist pilots in avoiding such areas when requested. Pilots should respond to a weather advisory by either acknowledging the advisory or by requesting an alternate course of action as appropriate.

Note: Remember that the controller's primary responsibility is to provide safe separation between aircraft. Any additional service, such as weather avoidance assistance, can only be provided to the extent that it does not derogate the primary function. ATC radar limitations and frequency congestion may also be a factor in limiting the controller's capability to provide additional service.

4. Give some examples of charts and reports useful in determining the potential for and location of thunderstorms along your route. (AC 00-45E)

a. A *convective outlook* (AC) describes the prospects for general thunderstorm activity during the following 24 hours. Areas in which there is a high, moderate, or slight risk of severe thunderstorms are included, as well as areas where thunderstorms may approach severe limits.

b. A *stability chart* outlines areas of stable and unstable air.

c. A *radar summary chart* graphically displays a collection of radar reports. The chart displays the type of precipitation echoes and indicates their intensity, intensity trend, configuration, coverage, echo tops and bases, and movement.

d. A *convective outlook chart* depicts areas of probable thunderstorm activity.

e. *Pilot Reports* (PIREPs) help in determining the actual conditions along your planned route of flight.

5. What are "microbursts"? (AIM 7-1-23)

Microbursts are small-scale intense downdrafts which, on reaching the surface, spread outward in all directions from the downdraft center. This causes the presence of both vertical and horizontal wind shears that can be extremely hazardous to all types and categories of aircraft, especially at low altitudes. Due to their small size, short life span, and the fact that they can occur over areas without surface precipitation, microbursts are not easily detectable using conventional weather radar or wind shear alert systems.

6. Where are microbursts most likely to occur? (AIM 7-1-23)

Microbursts can be found almost anywhere there is convective activity. They may be embedded in heavy rain associated with a thunderstorm or in light rain in benign-appearing virga. When there is little or no precipitation at the surface accompanying the microburst, a ring of blowing dust may be the only visual clue of its existence.

7. What are some basic characteristics of a microburst? (AIM 7-1-23)

Size: less than 1 mile in diameter as it descends from the cloud base; can extend 2½ miles in diameter near ground level.

Intensity: downdrafts as strong as 6,000 feet per minute; horizontal winds near the surface can be as strong as 45 knots resulting in a 90-knot wind shear (headwind to tailwind change for traversing aircraft).

Continued

Duration: an individual microburst will seldom last longer than 15 minutes from the time it strikes the ground until dissipation. Sometimes microbursts are concentrated into a line structure, and under these conditions activity may continue for as long as an hour.

8. How can microburst encounters be avoided? (AIM 7-1-23)

Pilots should heed windshear PIREPs, as a previous pilot's encounter with a microburst may be the only indication received. However, since the wind shear intensifies rapidly in its early stages, a PIREP may not indicate the current severity of a microburst.

9. Define the term "wind shear," and state the areas in which it is likely to occur. (AC 00-6A)

Wind shear is defined as the rate of change of wind velocity (direction and/or speed) per unit distance; conventionally expressed as vertical or horizontal wind shear. It may occur at any level in the atmosphere but three areas are of special concern:

a. Wind shear with a low-level temperature inversion;

b. Wind shear in a frontal zone or thunderstorm; and

c. Clear Air Turbulence (CAT) at high levels associated with a jet stream or strong circulation.

10. Why is wind shear an operational concern to pilots? (AC 00-6A)

Wind shear is an operational concern because unexpected changes in wind speed and direction can be potentially very hazardous to aircraft operations at low altitudes on approach to and departing from airports.

11. What airplane characteristics will be observed in the following wind shear situations?
   — a sudden increase in headwind.
   — a sudden decrease in headwind.

Increased headwind—As a tailwind shears to a constant head-wind, an increase in airspeed and altitude occurs along with a nose-up pitching tendency. The usual reaction is to reduce both power and pitch. This reaction can be dangerous if the aircraft suddenly encounters a downdraft and tailwind. Now the situation demands the exact opposite of the pilot's initial reaction: a need for more performance from the airplane instead of less (more power/increased pitch attitude).

Decreased headwind—As a headwind shears to a calm or tail-wind, pitch attitude decreases, airspeed decreases, and a loss of altitude occurs. The required action is more power and higher pitch attitude to continue a climb or remain on the glide slope.

12. Concerning wind shear detection, what does the abbreviation "LLWAS" indicate? (AIM 4-3-7)

Low-Level Wind Shear Alert System (LLWAS) is a computerized system that detects the presence of a possible hazardous low-level wind shear by continuously comparing the winds measured by sensors installed around the periphery of an airport with the wind measured at the center of the airport. If the difference between the center field wind sensor and a peripheral wind sensor becomes excessive, a thunderstorm or thunderstorm gust front wind shear is possible.

Additional Study Questions

1. When the ceiling/sky, visibility, and obstructions to vision are omitted in the ATIS, what condition is indicated? (AC 00-45E)

2. Give some examples of the different types of aviation weather charts available. (AC 00-45E)

3. Briefly describe the type of information provided by the different aviation weather charts. (AC 00-45E)

4. Define the terms IFR, MVFR, and VFR. (AC 00-45E)

5. Define the term "ceiling." (AC 00-45E)

6. From which primary source should information be obtained regarding expected weather at the ETA if your destination has no Aerodrome Forecast (TAF)? (AC 00-45E)

7. If the temperature is plus 16 degrees centigrade at an elevation of 1,600 feet and a standard (average) temperature lapse rate exists, what will the approximate freezing level be? (AC 00-6A)

8. What are the three stages of thunderstorm development? (AC 00-6A)

9. What is an "occlusion"? (AC 00-6A)

10. What information will you provide when giving a wind shear report? (AIM 7-1-21)

Airplane Systems

3

Some of the following questions are in reference to the systems of a Cessna 172-RG. For accuracy, a review of your aircraft's Pilot Operating Handbook should be made.

A. Primary Flight Controls and Trim

1. How are the various flight controls operated?

The flight control surfaces are manually actuated through use of either a rod or cable system. A control wheel actuates the ailerons and elevator, and rudder/brake pedals actuate the rudder.

2. What type of trim system is installed in this airplane?

Both rudder and elevator trim are provided. They are both manually actuated.

B. Wing Flaps, Leading Edge Devices, and Spoilers

1. What are flaps, and what is their function?
(FAA-H-8083-3)

The wing flaps are movable panels on the inboard trailing edges of the wings. They are hinged so they may be extended downward into the flow of air beneath the wings to increase both lift and drag. Their purpose is to permit a slower airspeed and a steeper angle of descent during a landing approach. In some cases, they may also be used to shorten the takeoff distance.

2. Describe a typical wing flap system.

The wing flap system consists of "single-slot" type wing flaps. They are extended and retracted by a wing flap switch lever to flap settings of 10, 20, and 30 degrees. A 15-amp push-to-reset circuit breaker protects the wing flap system circuit.

3. State some examples of leading edge lift devices.
(AC 25-14)

Slots—A slot in the leading edge of a wing directs high-energy air from under the wing to the airflow above the wing, accelerating upper airflow. By accelerating the airflow above the wing, airflow separation will be delayed to higher angles of attack. This allows the wing to continue to develop lift at substantially higher angles of attack.

Slats—A miniature airfoil mounted on the leading edge of a wing. They may be movable or fixed. At low angles of attack, movable slats are held flush against the leading edge by positive air pressure. At high angles of attack, the slats are moved forward either by the pilot or automatically by the low pressures present at the leading edge. Slats provide the same results as slots.

4. What are "spoilers"? (AC 65-15A)

Spoilers are devices located on the upper surface of a wing which are designed to reduce lift by "spoiling" the airflow above the wing. They are typically used as speed brakes to slow an airplane down, both in flight as well as on the ground immediately after touchdown.

C. Pitot Static System and Associated Flight Instruments

1. What instruments operate from the pitot/static system? (AC 61-27C)

The pitot/static system operates the altimeter, vertical speed indicator, and airspeed indicator.

2. Does this aircraft have an alternate static air system?

Yes, in the event of external static port blockage, a static pressure alternate source valve is installed. The control is located beneath the throttle, and if utilized will supply static pressure from inside the cabin, instead of from the external static ports.

3. How does an altimeter work? (AC 61-27C)

Aneroid wafers in the instrument expand and contract as atmospheric pressure changes, and through a shaft and gear linkage, rotate pointers on the dial of the instrument.

4. A pressure altimeter is subject to what limitations? (AC 61-27C)

Non-standard pressure and temperature:

a. Temperature variations expand or contract the atmosphere and raise or lower pressure levels that the altimeter senses.

On a warm day — The pressure level is higher than on a standard day. The altimeter indicates lower than actual altitude.

On a cold day — The pressure level is lower than on a standard day. The altimeter indicates higher than actual altitude.

b. Changes in surface pressure also affect pressure levels at altitude.

Higher than standard pressure — The pressure level is higher than on a standard day. The altimeter indicates lower than actual altitude.

Lower than standard pressure — The pressure level is lower than on a standard day. The altimeter indicates higher than actual altitude.

Remember: High to low or hot to cold, look out below!

5. Define and state how you would determine the following altitudes. (AC 61-27C)

Indicated altitude — Read off the face of the altimeter.

Pressure altitude — Indicated altitude with 29.92" Hg set in the Kollsman window.

True altitude — Height above sea level. Use the flight computer.

Density altitude — Pressure altitude corrected for non-standard temperature. Use the flight computer.

Absolute altitude — Height above ground. Subtract the terrain elevation from true altitude.

6. How does the airspeed indicator operate? (AC 61-27C)

It measures the difference between ram pressure from the pitot head and atmospheric pressure from the static source.

7. What are the limitations of the airspeed indicator? (AC 61-27C)

The airspeed indicator is subject to proper flow of air in the pitot/static system.

8. The airspeed indicator is subject to what errors? (AC 61-27C)

Position error—Caused by the static ports sensing erroneous static pressure; slipstream flow causes disturbances at the static port, preventing actual atmospheric pressure measurement. It varies with airspeed, altitude, and configuration, and may be a plus or minus value.

Density error— Changes in altitude and temperature are not compensated for by the instrument.

Compressibility error— Caused by the packing of air into the pitot tube at high airspeeds, resulting in higher than normal indications. It is usually not a factor.

9. What are the different types of aircraft speeds? (AC 61-27C)

Indicated airspeed — Read off instrument.

Calibrated airspeed — IAS corrected for instrument and position errors; obtained from the POH or off the face of instrument.

Equivalent airspeed — CAS corrected for adiabatic compressible flow at altitude.

True airspeed — EAS corrected for non-standard temperature and pressure; obtained from the flight computer, POH or A/S indicator slide computer. (Generally, EAS = CAS below 200 KTS and 10,000 feet.)

Ground speed — TAS corrected for wind; speed across ground; use the flight computer.

10. Are the color bands on an airspeed indicator indicated airspeeds or calibrated airspeeds? (AC 61-27C)

Airspeed indicators indicate calibrated airspeed (usually in mph) if manufactured in 1975 or before, and indicated airspeed (usually in knots) if manufactured in 1976 or after.

11. What airspeed limitations apply to the color-coded marking system of the airspeed indicator? (AC 61-23C)

white arc .... flap operating range

lower A/S limit white arc .... V_{S0} (stall speed landing configuration)

upper A/S limit white arc .... V_{FE} (maximum flap extension speed)

green arc .... normal operating range

lower A/S limit green arc .... V_S (stall speed clean or specified configuration)

upper A/S limit green arc .... V_{NO} (normal operations speed or maximum structural cruise speed)

yellow arc .... caution range (operations in smooth air only)

red line .... V_{NE} (maximum speed for operations in smooth air only)

12. What additional airspeed indicator markings are required in multi-engine airplanes?

14 CFR Part 23, which deals with Airworthiness Standards for airplanes of 12,500 pounds or less requires the following airspeed markings in multi-engine airplanes:

a. A blue radial line indicates best rate-of-climb airspeed with one engine inoperative (V_{YSE}).

b. A red radial line indicates the minimum controllable airspeed with one engine inoperative (V_{MC}).

13. How does the vertical speed indicator work? (AC 61-27C)

Changing pressures expand or contract a diaphragm connected to the indicating needle through gears and levers. The vertical speed indicator (VSI) is connected to the static pressure line through a calibrated leak; the VSI measures differential pressure.

14. What are the limitations of the vertical speed indicator? (AC 61-27C)

It is not accurate until the aircraft is stabilized. Sudden or abrupt changes in the aircraft attitude will cause erroneous instrument readings as airflow fluctuates over the static port. These changes are not reflected immediately by the VSI due to a calibrated leak.

D. Vacuum System and Associated Flight Instruments

1. What instruments contain gyroscopes? (AC 61-23C)

a. The turn and slip indicator/turn coordinator

b. The directional gyro (heading indicator)

c. Attitude indicator (artificial horizon)

2. What instruments operate from the vacuum system? (AC 61-27C)

Normally the attitude indicator and the directional gyro. The turn and slip/turn coordinator could also be vacuum-driven depending on the particular aircraft. The industry standard for aircraft dictates that the artificial horizon and directional gyro be vacuum-driven and the turn and slip/turn coordinator be electrically driven. However, in some systems all three can be electrically-driven.

3. How does the vacuum system operate? (AC 61-27C)

An engine-driven vacuum pump provides suction which pulls air from the instrument case. Normal pressure entering the case is directed against rotor vanes to turn the rotor (gyro) at high speed, much like a water wheel or turbine operates. Air is drawn into the instrument through a filter from the cockpit and eventually vented outside. Vacuum values vary but provide rotor speeds from 8,000 to 18,000 rpm.

4. How does the attitude indicator work? (AC 61-27C)

A gyro stabilizes the artificial horizon parallel to the real horizon.

5. What are the limitations of an attitude indicator?
(AC 61-27C)

The pitch and bank limits depend upon the make and model of the instrument. Limits in the banking plane are usually from 100 degrees to 110 degrees, and the pitch limits are usually from 60 to 70 degrees. If either limit is exceeded, the instrument will tumble or spill and will give incorrect indications until reset. A number of modern attitude indicators will not tumble.

6. The attitude indicator is subject to what errors?
(AC 61-27C)

Errors in both pitch and bank occur during normal coordinated turns. These errors are caused by the movement of pendulous vanes by centrifugal force resulting in precession of the gyro toward the inside of the turn. The greatest error occurs in 180° of turn. In a 180° turn to the right, on rollout the attitude indicator will indicate a slight climb and turn to the left. Acceleration and deceleration errors cause the attitude indicator to indicate a climb when the aircraft is accelerated and a descent when the aircraft is decelerated.

7. How does the directional gyro operate? (AC 61-27C)

A gyro stabilizes the heading indicator. The speed of the gyro is usually 10,000 to 18,000 rpm.

8. What are the limitations of the directional gyro?
(AC 61-27C)

The pitch and bank limits of the heading indicator vary with the particular design and make of instrument. On some heading indicators found in light airplanes, the limits are approximately 55 degrees of pitch and 55 degrees of bank. When either of these attitude limits is exceeded, the instrument "tumbles" or "spills" and no longer gives the correct indication until reset. After spilling, it may be reset with the caging knob. Many of the modern instruments used are designed in such a manner that they will not tumble.

9. The directional gyro is subject to what errors?
(AC 61-27C)

It is subject to precession of the gyro.

E. Electric/Gyroscopic System

1. What instruments operate on this system? (AC 61-27C)

Turn and Slip/Turn Coordinator — In some aircraft this system could also operate the artificial horizon and directional gyro (refer to vacuum/gyro system question #1 and #2, section D).

2. How does the turn and slip indicator/turn coordinator operate? (AC 61-27C)

A turn and slip indicator shows only the rate of turn while a turn coordinator shows both rate of roll and rate of turn. The turn part of the instrument uses precession to indicate direction and approximate rate of turn. A gyro reacts by trying to move in reaction to the force applied, thus moving the needle or miniature aircraft in proportion to the rate of turn. The slip/skid indicator is a liquid-filled tube with a ball that reacts to centrifugal force and gravity.

3. What information does the turn coordinator provide? (AC 61-27C)

The miniature aircraft of the turn coordinator displays the rate of turn and rate of roll. The ball in the tube indicates a slipping or skidding condition.

Slip — Ball on the inside of turn; not enough rate of turn for the amount of bank.

Skid — Ball to the outside of turn; too much rate of turn for the amount of bank.

4. What limitations apply to the turn coordinator? (AC 61-27C)

A spring is attached between the instrument case and the gyro assembly to hold the gyro upright when no precession force is applied. Tension on the spring may be adjusted to calibrate the instrument for a given rate of turn. The spring restricts the amount of gyro tilt. Stops prevent the gyro assembly from tilting more than 45 degrees to either side of the upright position.

F. Magnetic Compass

1. How does the magnetic compass work? (AC 61-27C)

Magnets mounted on the compass card align themselves parallel to the earth's lines of magnetic force.

2. What limitations does the magnetic compass have? (AC 61-27C)

The float assembly of the compass is balanced on a pivot, which allows free rotation of the card and allows it to tilt at an angle up to 18 degrees.

3. What are the various compass errors? (AC 61-27C)

Oscillation error— Erratic movement of the compass card caused by turbulence or rough control technique.

Deviation error—Due to electrical and magnetic disturbances in the aircraft.

Variation error— Angular difference between true and magnetic north; reference isogonic lines of variation.

Dip errors:

a. *Acceleration error*— On east or west headings, while accelerating, the magnetic compass shows a turn to the north, and when decelerating, it shows a turn to the south.

Remember: ANDS

A ccelerate

N orth

D ecelerate

S outh

Continued

b. *Northerly turning error*—The compass leads in the south half of a turn, and lags in the north half of a turn.

Remember: UNOS

U ndershoot

N orth

O vershoot

S outh

G. Hydraulic System

1. What equipment would be considered hydraulic on this aircraft?

a. The retractable landing gear

b. The emergency hand pump

c. The hydraulically-actuated brake on each main gear

d. The air/oil nose gear shock strut

2. What provides hydraulic power to the landing gear system?

An electrically-driven hydraulic power pack provides all hydraulic power to the landing gear system. The power pack is located behind the firewall between the pilot's and copilot's rudder pedals.

3. Describe hydraulic power pack operation.

Hydraulic power pack operation is controlled by the landing gear lever. When the gear lever is selected in either the "Up" or "Down" position, a pressure switch will activate the power pack and a selector valve is mechanically rotated. Depending on the position of the landing gear lever (and corresponding valve position), hydraulic pressure will be applied in the direction selected. This hydraulic pressure is applied to actuator cylinders, which extend or retract the gear. When the landing gear has reached the desired position and the cycle is complete (a series of electrical switches have closed or opened), an indicator light will illuminate on the panel. In the "Gear Down" cycle only, the hydraulic power pack will continue to operate until system pressure is between 1,000 PSI to 1,500 PSI, at which time the pressure switch turns the power pack off. The hydraulic system normally maintains an operating pressure of 1,000 PSI to 1,500 PSI.

H. Landing Gear

1. Describe the landing gear system on this airplane.

The landing gear consists of a tricycle-type system utilizing two main wheels and a steerable nose wheel. Tubular spring steel main gear struts provide main gear shock absorption, while nose gear shock absorption is provided by a combination air/oil shock strut.

2. How is the landing gear extended and retracted?

A hydraulic actuator powered by an electrically-driven hydraulic power pack enables the landing gear extension, retraction, and main gear down lock release operations to occur. A pressure switch starts and stops power pack operation and hydraulic pressure is directed by a landing gear lever.

3. How is the gear locked in the down position?

Mechanical down locks are incorporated into the nose and main gear assembly.

4. How is the gear locked in the up position?

A positive "up" pressure is maintained on the landing gear by the hydraulic power pack. To accomplish this, the power pack automatically maintains an operating pressure of 1,000 PSI to 1,500 PSI in the landing gear system.

5. How is accidental gear retraction prevented on the ground?

Inadvertent gear retraction is prevented by a safety (squat) switch on the nose gear. Whenever the nose gear strut is compressed (weight of the airplane on the ground), this switch electrically prevents operation of the landing gear system.

6. How is the landing gear position indicated in the cockpit?

Amber (gear up) and green (gear down) position indicator lights are provided in the cockpit. They are located adjacent to the landing gear control lever and indicate that the gear is either up or down and locked. Both indicators incorporate a press-to-test feature and also provide dimming shutters for night operation.

Note: If one of the indicator lights should burn out, the design allows for replacement inflight, with the bulb from the other indicator light.

7. What type of landing gear warning system is used?

If the manifold pressure is reduced to less than approximately 12 inches at a low altitude with the master switch on, and the landing gear is not locked down, a switch on the throttle linkage will electrically actuate the gear warning circuit of the dual warning unit. An intermittent tone will be heard on the speaker. Also, if the wing flaps are extended beyond 20° while the landing gear is in the retracted position, an interconnect switch in the wing flap system will cause the horn to sound.

8. What is the normal length of time necessary for landing gear retraction or extension?

5 to 7 seconds for either extension or retraction of the landing gear.

9. Can the landing gear be retracted with the hand-operated pump?

No, retraction of the landing gear cannot be accomplished with the emergency hand pump.

10. Describe the braking system on this aircraft.

Hydraulically-actuated disc-type brakes are utilized on each main gear wheel. A hydraulic line connects each brake to a master cylinder located on each pilot's rudder pedals. By applying pressure to the top of either the pilot's or copilot's set of rudder pedals, the brakes may be applied.

11. How is steering accomplished on the ground?
(AC 61-23C)

Light airplanes are generally provided with nosewheel steering capabilities through a simple system of mechanical linkages connected to the rudder pedals. When a rudder pedal is depressed, a spring-loaded bungee (push-pull rod) connected to the pivotal portion of a nosewheel strut will turn the nosewheel.

12. What are the landing gear tire pressures?

Nosewheel Tire Pressure .... 40 – 50 PSI
                       5.00-5, 6-ply rated tires

Main Wheel Tire Pressure ..... 60 – 68 PSI
                       15x6.00-6, 6-ply rated tires

I. Powerplant

1. What type of engine does this aircraft have?

The airplane is powered by an engine manufactured by Avco-Lycoming, rated at 180 horsepower at 2,700 rpm. It may be described as follows:

a. Normally aspirated
b. Direct-drive
c. Air-cooled
d. Horizontally opposed
e. Carburetor-equipped
f. Four-cylinder
g. 361-cubic-inch displacement

2. Describe how each of the following engine gauges work.

Oil Temperature—Electrically powered from the aircraft electrical system.

Oil Pressure—A direct-pressure oil line from the engine delivers oil at engine operating pressure to the gauge.

Cylinder Head Temperature—Electrically powered from the aircraft electrical system.

Continued

Tachometer—Engine-driven mechanically.

Manifold pressure—Direct reading of induction air manifold pressure in inches of mercury.

Fuel pressure—Indicates fuel pressure to the carburetor.

3. What series of operations or events must occur in each cylinder of a typical four-stroke reciprocating engine in order for it to produce full power? (AC 61-23C)

The events that occur are:

a. Intake

b. Compression

c. Power

d. Exhaust

This cycle requires four strokes of the piston, two up and two down. Ignition of the fuel/air mixture at the end of the compression stroke adds a fifth event; consequently, the cycle of events is known as the four-stroke, five-event cycle principle.

4. What does the carburetor do? (AC 61-23C)

Carburetion may be defined as the process of mixing fuel and air in the correct proportions to form a combustible mixture. The carburetor vaporizes liquid fuel into small particles and then mixes it with air. It measures the airflow and meters fuel accordingly.

5. How does the carburetor heat system work?

A carburetor heat valve, controlled by the pilot, allows unfiltered, heated air from a shroud located around an exhaust riser or muffler to be directed to the induction air manifold prior to the carburetor. Carburetor heat should be used anytime suspected or known carburetor icing conditions exist.

6. What is "fuel injection"? (FAA-H-8083-3)

Fuel injectors have replaced carburetors in some airplanes. In a fuel injection system, the fuel is normally injected into the system either directly into the cylinders or just ahead of the intake valves;

whereas in a carbureted system, the fuel enters the airstream at the throttle valve. There are several types of fuel injection systems in use today, and though there are variations in design, the operational methods are generally simple. Most designs incorporate an engine-driven fuel pump, a fuel/control unit, fuel distributor, and discharge nozzle for each cylinder.

7. What are some advantages of fuel injection? (AC 61-23C)

a. Reduction in evaporative icing

b. Better fuel flow

c. Faster throttle response

d. Precise control of mixture

e. Better fuel distribution

f. Easier cold weather starts

8. Are there any disadvantages associated with fuel-injected engines? (AC 61-23C)

a. Difficulty in starting a hot engine

b. Vapor locks during ground operations on hot days

c. Problems associated with restarting an engine that quits because of fuel starvation

9. What is the condition known as "vapor lock"? (AC 61-23C)

Fuel-injected engines are more susceptible to vapor lock during ground operations on hot days. Vapor lock occurs when fuel vapor and/or air collect in different sections of the fuel system. The fuel may actually boil in the lines, creating a condition which interferes with the normal operation of valves, pumps, etc. Other causes of vapor lock can be low fuel pressure and excessive fuel turbulence.

10. What does the throttle do? (FAA-H-8083-3)

The throttle allows the pilot to manually control the amount of fuel/air charge entering the cylinders. This in turn regulates the engine manifold pressure.

11. What does the mixture control do? (FAA-H-8083-3)

It regulates the fuel-to-air ratio. Most airplane engines incorporate a device called a mixture control, by which the fuel/air ratio can be controlled by the pilot during flight. The purpose of a mixture control is to prevent the mixture from becoming too rich at high altitudes, due to decreasing air density. Leaning the mixture during cross-country flights conserves fuel and provides optimum power.

12. What are "turbochargers"? (AC 65-12A)

Higher performance aircraft typically operate at higher altitudes where air density is substantially less. The decrease in air density as altitude increases results in a decreased power output of an unsupercharged engine. By compressing the thin air by means of a supercharger, the turbocharged engine will maintain the preset power as altitude is increased. The turbocharger consists of a compressor to provide pressurized air to the engine, and a turbine driven by exhaust gases of the engine to drive the compressor.

13. What are "cowl flaps"?

Cowl flaps are located on the engine cowling and allow the pilot to control the operating temperature of the engine by regulating the amount of air circulating within the engine compartment. Cowl flaps may be manually or electrically activated and usually allow for a variety of flap positions.

14. When are cowl flaps utilized?

a. Normally the cowl flaps will be in the "open" position in the following operations:
   i. During starting of the engine
   ii. While taxiing
   iii. During takeoff and high-power climb operation

The cowl flaps may be adjusted in cruise flight for the appropriate cylinder head temperature.

b. The cowl flaps should be in the "closed" position in the following operations:

    i. During extended let-downs

    ii. Anytime excessive cooling is a possibility (i.e., approach to landing, engine-out practice, etc.)

15. Define the term "service ceiling."

Service Ceiling—Defined as the height above sea level beyond which the aircraft's maximum rate of climb would be no more than 100 feet per minute. The service ceiling may be found in the Pilot's Operating Handbook for that aircraft.

J. Propeller

1. What type of propeller does this aircraft have?

The airplane propeller may be described as

a. All-metal,

b. Two-bladed,

c. Constant-speed, and

d. Governor-regulated.

2. Discuss fixed-pitch propellers. (FAA-H-8083-3)

An airplane equipped with a fixed-pitch propeller is designed for best efficiency at only one rotational speed (rpm) and one forward speed (airspeed). Any change in these conditions reduces the efficiency of both the propeller and the engine.

3. Discuss variable-pitch propellers (constant speed). (FAA-H-8083-3)

An airplane equipped with a constant-speed propeller is capable of continuously adjusting the propeller blade angle to maintain a constant engine speed. For example, if engine rpm increases as a result of a decreased load on the engine (descent), the system automatically increases the propeller blade angle (increasing air

Continued

load) until the rpm has returned to the preset speed. The propeller governor can be regulated by the pilot with a control in the cockpit, so that any desired blade angle setting (within its limits) and engine operating rpm can be obtained, thereby increasing the airplane's efficiency in various flight conditions.

4. What does the propeller control do? (FAA-H-8083-3)

The propeller control regulates propeller pitch and engine rpm as desired for a given flight condition. The propeller control adjusts a propeller governor which establishes and maintains the propeller speed, which in turn maintains the engine speed.

5. What would the desired propeller setting be for maximum performance situations such as takeoff? (FAA-H-8083-3)

A low pitch, high rpm setting produces maximum power and thrust. The low blade angle keeps the angle of attack small and efficient with respect to the relative wind. At the same time, it allows the propeller to handle a smaller mass of air per revolution. This light load allows the engine to turn at high rpm and to convert the maximum amount of fuel into heat energy in a given time. The high rpm also creates maximum thrust because the mass of air handled per revolution is small, the number of revolutions per minute is many, the slipstream velocity is high, and the airplane speed is low.

6. What is a propeller governor?

The propeller governor, with the assistance of a governor pump, controls the flow of engine oil to or from a piston in the propeller hub. When the engine oil, under high pressure from the governor pump, pushes the piston forward, the propeller blades are twisted toward a high pitch/low rpm condition. When the engine oil is released from the cylinder, centrifugal force, with the assistance of an internal spring, twists the blades towards a low pitch/high rpm condition.

7. **When operating an airplane with a constant-speed propeller, which condition induces the most stress on the engine?** (FAA-H-8083-3)

High manifold pressure/low rpm. If an excessive amount of manifold pressure is carried for a given rpm, the maximum allowable pressure within the engine cylinders could be exceeded, placing undue stress on them. If repeated too frequently, this undue stress could weaken the cylinder components and eventually cause engine structural failure.

K. Fuel System

1. What type fuel system does this aircraft have?

The fuel system consists of the following:

a. Two vented integral fuel tanks
b. A four-position fuel selector valve
c. A fuel strainer
d. A manual primer
e. An engine-driven fuel pump
f. An electric auxiliary fuel pump
g. A carburetor

The airplane utilizes a gravity-feed type fuel system. Fuel is delivered to the engine-driven fuel pump, unassisted, except by gravity. Fuel flows from the wing tanks to the fuel selector valve which is marked with BOTH, RIGHT, LEFT and OFF positions. From the fuel valve, fuel flows through a fuel strainer and then to the engine-driven fuel pump. The fuel pump then delivers fuel to the carburetor. After the carburetor, the fuel/air mixture is delivered to the cylinders via intake manifold tubes.

2. When is the auxiliary fuel boost pump used?

Proper usage of the fuel boost pump varies with different aircraft. In general, the fuel boost pump should be used during takeoffs and landings, when switching fuel tanks, and anytime fuel pressure falls below a selected value. The fuel boost pump in a Cessna 172-RG should be utilized anytime the fuel pressure falls below 0.5 PSI.

3. Why is it necessary to include a left and right position on the fuel selector valve?

During cruise flight, with the fuel selector valve on "Both," unequal fuel flow may occur if the wings are not consistently kept level during the flight. This will result in one wing being heavier than the other. A fuel selector valve with the left/right option allows a pilot to control the situation by selecting the tank on the heavier wing and remaining on that tank until both tanks contain approximately the same amount of fuel.

4. Where are the fuel vents located for each tank?

The left fuel tank is vented overboard through a vent line with a check valve. The right fuel tank is vented through the filler cap. Both fuel tanks are vented together by an interconnecting line.

5. What purpose do fuel tank vents have? (FAA-H-8083-3)

As the fuel level in an aircraft fuel tank decreases, without vents a vacuum would be created within the tank which would eventually result in a decreasing fuel flow and finally engine stoppage. Fuel system venting provides a way of replacing fuel with outside air, preventing formation of a vacuum.

6. What type fuel does this aircraft require (minimum octane rating and color)?

The approved fuel grade used is 100LL, and the color is blue.

7. Can other types of fuel be used if the specified grade is not available? (AC 91-33A)

You may use fuel of a higher grade but only as a temporary solution. In no case should you ever use a fuel of a lower grade such as 80/87. If you must use a different grade of fuel, use a grade as close as possible to 100LL such as 100/130 or 115/145, and use it only for a short period of time.

8. What are some examples of different fuel grades (octane ratings) available?

Grade (octane)   Color (dye)

80 .... Red
100 .... Green
100LL .... Blue
115 .... Purple
Turbine .... Colorless

9. What is the function of the manual primer, and how does it operate?

The manual primer's function is to provide assistance in starting the engine. The primer draws fuel from the fuel strainer and injects it directly into the cylinder intake ports. This usually results in a quicker, more efficient engine start.

10. Where are the drain valves located?

A drain valve is located on the bottom of each main wing and also directly under the fuel selector valve. A fuel strainer drain is located under an access panel on the right side of the engine cowling.

11. How is fuel quantity measured?

One float-type fuel quantity transmitter and one electric fuel quantity indicator measure fuel quantity for each tank.

12. Are the fuel quantity indicators accurate?

In general, the indicators in most aircraft should not be relied on completely for fuel management. Also, the fuel indicators are not accurate during skids, slips, or unusual attitudes.

L. Oil System

1. Briefly describe the engine oil system.

Aircraft engine lubrication and oil for propeller governor operation is supplied from a sump on the bottom of the engine. Oil sump capacity is 8 quarts.

2. What are the minimum and maximum oil capacities?

The minimum oil capacity is 5 quarts of oil. The maximum oil capacity is 8 quarts.

3. What are the minimum and maximum oil temperatures and pressures?

Oil temperature—100°F to 245°F

Oil pressure—25 PSI (minimum for idling), 60 – 90 PSI (green arc), 115 PSI (red line)

4. What are two types of oil available for use in your airplane?

Mineral oil—Also known as non-detergent oil; contains no additives. This type of oil is normally used after an engine over-haul or when an aircraft engine is new; normally used for engine break-in purposes.

Ashless dispersant—Mineral oil with additives; high antiwear properties along with multi-viscosity (ability to perform in wide range of temps). Also picks up contamination and carbon particles and keeps them suspended so that buildups and sludge do not form in the engine.

5. What type of oil is recommended for this engine (for summer and winter operations)?

Ashless dispersant oil, usually SAE 20W-50 during the colder months. For temperatures above 60°F (summer), use SAE 40 or SAE 50.

M. Electrical System

1. Describe the electrical system on this aircraft.

Electrical energy is provided by a 28-volt, direct-current system, powered by an engine-driven 60-amp alternator and a 24-volt battery.

2. Where is the battery located?

The battery is located aft of the rear cabin wall.

3. How are the circuits for the various electrical accessories within the aircraft protected?

Most of the electrical circuits in an airplane are protected from an overload condition by either circuit breakers or fuses or both. Circuit breakers perform the same function as fuses except that when an overload occurs, a circuit breaker can be reset.

4. What is a "bus bar"?

A "bus bar" interfaces the electrical system with the various electrical accessories such as radios, lights, etc. A bus bar makes aircraft electrical wiring considerably less complex. Fuses or circuit breakers are incorporated into the bus bar providing protection for the different aircraft electrical circuits from possible damage as a result of faulty generators, alternators or components.

5. The electrical system provides power for what equipment in an airplane?

Normally the following:

a. Radio equipment	h. Interior lights
b. Turn coordinator	i. Instrument lights
c. Fuel gauges	j. Position lights
d. Pitot heat	k. Flaps (maybe)
e. Landing light	l. Stall warning system (maybe)
f. Taxi light	m. Oil temperature gauge
g. Strobe lights	n. Cigarette lighter (maybe)

6. What does the ammeter indicate?

The ammeter indicates the flow of current, in amperes, from the alternator to the battery or from the battery to the electrical system. With the engine running and master switch on, the ammeter will indicate the charging rate to the battery. If the alternator has gone off-line and is no longer functioning, or the electrical load exceeds the output of the alternator, the ammeter indicates the discharge rate of the battery.

7. What function does the voltage regulator have?

The voltage regulator is a device which monitors system voltage, detects changes, and makes the required adjustments in the output of the alternator to maintain a constant regulated system voltage. It must do this at low rpms, such as during taxi, as well as at high rpms in flight. In a 28-volt system, it will maintain 28 volts ±0.5 volts.

8. Does the aircraft have an external power source receptacle, and if so where is it located?

Yes, the receptacle is located behind a door on the left side of the fuselage aft of the baggage compartment door.

9. What type of ignition system does your airplane have? (AC 61-23C)

Engine ignition is provided by two engine-driven magnetos, and two spark plugs per cylinder. The ignition system is completely independent of the aircraft electrical system. The magnetos are self-contained units supplying electrical current without using an external source of power. However, before they can produce current, the magnetos must be actuated as the engine crankshaft is rotated by some other means. To accomplish this, the aircraft battery furnishes electrical power to operate a starter which, through a series of gears, rotates the engine crankshaft. This in turn actuates the armature of the magneto to produce the sparks for ignition of the fuel in each cylinder. After the engine starts, the starter system is disengaged and the battery no longer contributes to the actual operation of the engine.

10. **What are the two main advantages of a dual ignition system?** (AC 61-23C)

    a. Increased safety — in case one system fails the engine may be operated on the other until a landing is safely made.

    b. More complete and even combustion of the mixture, and consequently improved engine performance; i.e., the fuel/air mixture will be ignited on each side of the combustion chamber and burn toward the center.

N. Environmental System

1. How does the aircraft cabin heat work?

Fresh air, heated by an exhaust shroud, is directed to the cabin through a series of ducts.

2. How does the pilot control temperature in the cabin?

Temperature is controlled by mixing outside air (cabin air control) with heated air (cabin heat control) in a manifold near the cabin firewall. This air is then ducted to vents located on the cabin floor.

3. What are the two types of oxygen breathing systems normally in use? (AC 65-15A)

The type of oxygen system used is determined by the regulator; the two types are continuous-flow and pressure-demand. Most general aviation aircraft use the continuous-flow type.

4. Can any kind of oxygen be used for aviator's breathing oxygen?

No, oxygen used for medical purposes or welding normally should not be used because it may contain too much water. The excess water could condense and freeze in the oxygen lines when flying at high altitudes. Specifications for "aviator's breathing oxygen" are 99.5% pure oxygen with not more than two milliliters of water per liter of oxygen.

5. How does a continuous-flow oxygen system operate?

In a continuous-flow system the amount of oxygen delivered is controlled by a calibrated orifice. It is usually characterized by the bag attached to the mask. Oxygen is continuously dispensed from the oxygen storage bottle through the calibrated orifice to the bag and mask. As the wearer inhales the oxygen from the rebreather bag, it deflates; as the wearer exhales, some of the unused oxygen is forced back into the bag and mixed with 100% oxygen for the next inhalation. Some of the oxygen is forced out through a cluster of small orifices, either in the nose of the mask or on either side. To check for oxygen flow, there is usually a mechanical oxygen flow indicator (of some color) in the tube leading to the rebreather bag.

6. What is a "pressurized" aircraft?

In a "pressurized" aircraft, the cabin, flight compartment, and baggage compartments are incorporated into a sealed unit which is capable of containing air under a pressure higher than outside atmospheric pressure. Pressurized air is pumped into this sealed fuselage by cabin superchargers which deliver a relatively constant volume of air at all altitudes up to a designed maximum. Air is released from the fuselage by a device called an outflow valve. Since the superchargers provide a constant inflow of air to the pressurized area, the outflow valve, by regulating the air exit, is the major controlling element in the pressurization system.

7. What operational advantages are there in flying pressurized aircraft? (FAA-H-8083-3)

A cabin pressurization system performs several functions:

a. It allows an aircraft to fly higher which can result in better fuel economy, higher speeds, and the capability to avoid bad weather and turbulence.

b. It will typically maintain a cabin pressure altitude of 8,000 feet at the maximum designed cruising altitude of the airplane.

c. It prevents rapid changes of cabin altitude which may be uncomfortable or injurious to passengers and crew.

d. It permits a reasonably fast exchange of air from inside to outside of the cabin. This is necessary to eliminate odors and to remove stale air.

8. Describe a typical cabin pressure control system. (FAA-H-8083-3)

The cabin pressure control system provides cabin pressure regulation, pressure relief, vacuum relief, and the means for selecting the desired cabin altitude in the isobaric and differential range. In addition, dumping of the cabin pressure is a function of the pressure control system. A cabin pressure regulator, an outflow valve, and a safety valve are used to accomplish these functions.

9. What are the components of a cabin pressure control system? (FAA-H-8083-3)

a. *Cabin pressure regulator*—Controls cabin pressure to a selected value in the isobaric range and limits cabin pressure to a preset differential value in the differential range.

b. *Cabin air pressure safety valve*— A combination pressure relief, vacuum relief, and dump valve.

   i. Pressure relief valve: prevents cabin pressure from exceeding a predetermined differential pressure above ambient pressure.

   ii. Vacuum relief valve: prevents ambient pressure from exceeding cabin pressure by allowing external air to enter the cabin when the ambient pressure exceeds cabin pressure.

   iii. Dump valve: actuated by a cockpit control which will cause the cabin air to be dumped to the atmosphere.

O. Ice Prevention and Elimination

1. What is the difference between a deice system and an anti-ice system? (AC 65-15A)

A deice system is used to eliminate ice that has already formed. An anti-ice system is used to prevent the formation of ice.

2. What types of systems are utilized in the prevention and elimination of airframe ice? (AC 65-15A)

Pneumatic — A deice type of system; consists of inflatable boots attached to the leading edges of the wings and tail surfaces. Compressed air from the pressure side of the engine vacuum pump is cycled through ducts or tubes in the boots causing the boots to inflate. Most systems also incorporate a timer.

Hot Air — An anti-ice type system; commonly found on turboprop and turbojet aircraft. Hot air is directed from the engine (compressor) to the leading edges of the wings.

3. What types of systems are utilized in the prevention and elimination of propeller ice? (AC 65-12A)

Electrically heated boots — Consist of heating elements incorporated into the boots which are bonded to the propeller. The ice buildup on the propeller is heated from below and then thrown off by centrifugal force.

Fluid system — Consists of an electrically driven pump which, when activated, supplies a fluid, such as alcohol, to a device in the propeller spinner which distributes the fluid along the propeller assisted by centrifugal force.

4. What types of systems are utilized in the prevention and elimination of windshield ice? (AC 65-15A)

Fluid system — Consists of an electrically-driven pump which may be activated to spray a fluid, such as alcohol, onto the windshield to prevent the formation of ice.

Electrical system—Heating elements are embedded in the wind-shield or in a device attached to the windshield which when activated, prevents the formation of ice.

P. Avionics

1. What function does the avionics power switch have?

The avionics power switch controls power from the primary bus to the avionics bus. The circuit is protected by a combination power switch/circuit breaker. Aircraft avionics are isolated from electrical power when the switch is in the "Off" position. Also, if an over-load should occur in the system, the avionics power switch will move to the "Off" position, causing an interruption of power to all aircraft avionics.

2. What are static dischargers? (AC 00-6A)

If frequent IFR flights are planned, installation of wick-type static dischargers is recommended to improve radio communications during flight through dust or various forms of precipitation (rain, snow or ice crystals). Under these conditions, the buildup and discharge of static electricity from the trailing edges of the wings, rudder, elevator, propeller tips, and radio antennas can result in loss of usable radio signals on all communications and navigation radio equipment. Usually the ADF is the first to be affected and VHF communication equipment is the last to be affected.

3. Within what frequency band does the following type of navigational and communication equipment installed on board most aircraft operate? (AIM)

VOR receiver (VHF band) .... 108.0 to 117.95 MHz

Communication transceivers (VHF band)... 118.0 to 135 MHz

DME receiver (UHF band) .... 962 MHz to 1213 MHz

ADF receiver (LF to MF band) .... 190 to 530 kHz

Note: Know the location and function of the various antennas on your aircraft!

Additional Study Questions

1. During a preflight runup, the pilot switches from both magnetos to the left magneto and notes there is no drop in rpm. What does this indicate? (AC 61-23C)

2. A pilot notices that when applying carburetor heat during an engine runup on the ground, an increase in rpm occurs. What does this indicate? (AC 61-23C)

3. What is the purpose of "differential" aileron travel? (FAA-H-8083-3)

4. The attitude indicator contains several instrument markings. What information do they provide? (FAA-H-8083-3)

5. During gear extension or retraction, the electric hydraulic pump malfunctions and continues to run after the gear has been extended or retracted. How would the pilot detect this malfunction? (POH)

6. Where is the hydraulic reservoir located on your aircraft? What type fluid does it use? (POH)

7. What indications will you have if a loss of hydraulic pressure occurs in flight? (POH)

8. What is the maximum number of degrees the nose wheel will travel left or right of center? (POH)

9. Where are the brake system master cylinder(s) located on your aircraft? (POH)

10. If a loss of oil pressure in the propeller occurs, what effect will this have on the propeller pitch and rpm? (POH)

Emergency
Procedures

Some of the following questions are in reference to the systems of a Cessna 172-RG. For accuracy, a review of your aircraft's Pilot Operating Handbook should be made.

A. Spin Recovery

1. What is the recommended procedure for recovery from a spin? (FAA-H-8083-3)

Anytime a spin is encountered, regardless of the conditions, the normal spin recovery should be used.

a. Retard the throttle to idle.
b. Apply full opposite rudder to slow the rotation.
c. Apply full positive forward elevator movement to break the stall.
d. Neutralize the rudder as the spinning stops.
e. Return to level flight.

2. What does an aft center of gravity do to an aircraft's spin characteristics? (FAA-H-8083-3)

Recovery from a stall in any aircraft becomes progressively more difficult as its center of gravity moves aft. This is particularly important in spin recovery, as there is a point in rearward loading of any airplane at which a "flat" spin will develop. A "flat" spin is one in which centrifugal force acting through a center of gravity located well to the rear, will pull the tail of the airplane out away from the axis of the spin, making it impossible to get the nose down and recover.

3. What load factor is present in a spin? (FAA-H-8083-3)

The load factor during a spin will vary with the spin characteristics of each airplane but is usually found to be slightly above the 1G load of level flight. There are two reasons this is true:

a. The airspeed in a spin is very low (usually within 2 knots of the unaccelerated stalling speed); and
b. The airplane pivots, rather than turns, while it is in a spin.

B. Emergency Checklist

Discuss the use of an "emergency checklist."

In the event of an in-flight emergency, the pilot should be sufficiently familiar with emergency procedures to take immediate action instinctively to prevent more serious situations from occurring. However, as soon as circumstances permit, the emergency checklist should be reviewed to ensure that all required items have been checked. Additionally, before takeoff, a pilot should be sure that the emergency checklist will be readily accessible in flight if needed.

C. Partial Power Loss

What procedures should be followed concerning a partial loss of power in flight?

If a partial loss of power occurs, the first priority is to establish and maintain a suitable airspeed (best glide airspeed if necessary).

Then, select an emergency landing area and remain within gliding distance. As time allows, attempt to determine the cause and correct it. Complete the following checklist:

a. Check the carburetor heat.

b. Check the amount of fuel in each tank and switch fuel tanks if necessary.

c. Check the fuel selector valve's current position.

d. Check the mixture control.

e. Check that the primer control is all the way in and locked.

f. Check the operation of the magnetos in all three positions: both, left or right.

D. Engine Failure

1. In the event of a complete engine failure on takeoff, what procedure is recommended?

If an engine failure occurs during the takeoff run, the following checklist should be completed:

a. Retard the throttle to idle.

b. Apply pressure to the brakes.

c. Retract the wing flaps.

d. Set the mixture control to "Idle Cut-off."

e. Turn the ignition switch to "Off."

f. Turn the master switch to "Off."

2. If an engine failure occurs immediately after takeoff, what procedure is recommended? (FAA-H-8083-3)

If an engine failure occurs immediately after takeoff, and before a safe maneuvering altitude is attained, it is usually inadvisable to attempt to turn back to the field from which the takeoff was made. Instead, it is generally safer to immediately establish the proper glide attitude, and select a field directly ahead or slightly to either side of the takeoff path.

The following checklist should be completed:

a. Set the mixture control to "Idle Cutoff."

b. Set the fuel selector valve to "Off."

c. Turn the master switch to "Off."

d. Turn the ignition switch to "Off."

e. Establish an airspeed with flaps up of 75 KIAS, and with flaps down, of 65 KIAS.

f. Set the wing flaps as required (30° recommended).

3. What is the recommended procedure to be followed for an engine failure while en route?

If an engine failure occurs while en route, the first priority is to establish a best-glide airspeed. Then, select an emergency landing area and remain within gliding distance. As time permits, try to determine the cause of the failure (no fuel, carburetor ice, etc.). Attempt an engine restart if possible.

The following recommended checklist should be completed during the procedure:

a. Establish an airspeed of 75 KIAS.

b. Set the mixture control to "Rich."

Continued

c. Set the carburetor heat to "On."

d. Set the fuel selector valve to "Both."

e. Turn the ignition switch to "Both"; if the propeller has stopped, turn the ignition switch to "Start."

f. Check that the primer control is pushed in and is "locked."

4. What is the recommended power-off gliding speed in an engine-out procedure?

73 KIAS at 2,650 pounds.

E. Emergency Landing

1. If an engine failure has occurred while en route and a forced landing is imminent, what procedures should be followed? (FAA-H-8083-3)

a. Establish an airspeed of 75 KIAS.

b. Begin a scan for an appropriate field for landing using the following order of preference:

   i. Paved airport
   ii. Unpaved airport
   iii. Paved road with no obstacles
   iv. Unpaved road with no obstacles
   v. Grass field
   vi. Plowed field
   vii. Lakes or ponds
   viii. Trees or other structures

c. Attempt an engine restart.

d. Set your transponder to "7700."

e. Transmit a "mayday" message on either the frequency in use or 121.5.

f. Begin to spiral down over the approach end of the selected landing site.

g. On your final approach complete the forced landing checklist.

2. Immediately before touchdown in a forced landing procedure, what items should be completed?

The Emergency Landing Checklist should be completed:

a. Set the mixture control to "Idle Cut-Off."

b. Set the wing flaps as required (30° recommended).

c. Select the landing gear down or up depending on terrain.

d. Make sure all doors are unlatched prior to touchdown.

e. Set the fuel selector valve to "Off."

f. Establish an airspeed with flaps up of 75 KIAS and with flaps down of 65 KIAS.

g. Turn the ignition switch to "Off."

h. Turn the master switch to "Off."

i. Make your touchdown with the tail slightly low.

j. Apply brakes as needed.

3. In an engine failure situation, what glide ratio will be obtained if the best-glide airspeed is maintained?

A loss of 600 feet per 1 nautical mile (i.e., an aircraft at 3,000 feet AGL would have a maximum gliding distance of 5 miles).

4. If a forced landing is imminent, should the landing gear be left up, or down and locked?

There is no established set of rules concerning the positioning of the retractable landing gear in forced landing situations. Always follow the manufacturer's recommendation. In general, if the selected field is relatively smooth and hard surfaced, the landing gear should normally be left down. If the field is unimproved (rough terrain or a soft landing surface), the gear should be left up to minimize damage and improve directional control during the landing.

Note: In some cases (high descent rate, extremely rough terrain, etc.) an extended landing gear may minimize impact to the cockpit or cabin area.

5. If an engine failure has occurred while over water, and you are beyond power-off gliding distance to land, what procedures should be followed? (FAA-H-8083-3)

a. Establish an airspeed of 75 KIAS.

b. Attempt an engine restart.

c. Set your transponder to "7700" and broadcast a "mayday" message on the frequency in use or 121.5.

d. Make sure all heavy objects are secured or, if possible, jettison them.

e. Locate the life raft and life vests. (Don vests, time permitting.)

f. Approach and land parallel to heavy sea swells when in light winds and approach and land into the wind when high winds and heavy seas exist.

g. Set flaps to 10 degrees and establish 65 KIAS.

h. Open all cabin doors before touchdown.

i. Protect your body with life vests, clothing, etc.

j. Set up a minimum descent rate.

k. Initiate your touchdown in a level flight attitude.

l. After touchdown, begin evacuation of the airplane. Open the windows to equalize pressure if the doors do not open easily.

F. Engine Roughness or Overheat

1. What is "detonation"? (AC 61-23C)

Detonation is caused when the fuel/air mixture is subjected to a combination of excessively high temperatures and high pressure within the cylinder, resulting in an instantaneous burning (explosion) of the mixture. This form of combustion causes a definite loss of power, engine overheating, preignition and, if allowed to continue, physical damage to the engine. It may be caused by the wrong grade of fuel.

2. What action should be taken if detonation is suspected? (AC 61-23C)

Corrective action for detonation may be accomplished by adjusting any of the engine controls which will reduce both temperature and pressure of the fuel air charge.

a. Open cowl flaps (if installed);
b. Reduce power;
c. Reduce climb rate (increases air flow);
d. Enrich fuel/air mixture.

3. What is "preignition"? (AC 61-23C)

Preignition is defined as ignition of the fuel prior to normal ignition, or ignition before the electrical arcing occurs at the spark plugs. Preignition may be caused by excessively hot exhaust valves, or carbon particles on spark plug electrodes heated to an incandescent or glowing state. In most cases these local "hot spots" are caused by the high temperatures encountered during detonation.

4. What actions should be taken if preignition is suspected? (AC 61-23C)

Corrective actions for preignition include any type of engine operation which would promote cooling such as:

a. Open cowl flaps (if installed);
b. Reduce power;
c. Reduce climb rate (increases air flow);
d. Enrich fuel/air mixture.

5. If the engine begins to run rough when flying through heavy rain, what action should be taken?

During flight through heavy rain, it is possible for the induction air filter to become water saturated. This situation will reduce the amount of available air to the carburetor resulting in an excessively rich mixture and a corresponding loss of power. Carburetor heat may be used as an alternate source of air in such a situation.

6. Are there any special considerations necessary when using the auxiliary pump after an engine-driven fuel pump failure?

In a high-wing, single-engine aircraft, which has sustained an engine-driven fuel pump failure, gravity flow will provide sufficient fuel flow for level or descending flight. If the failure occurs while in a climb or the fuel pressure falls below 0.5 PSI, the auxiliary fuel pump should be utilized.

7. What operating procedure could be used to minimize spark plug fouling?

Engine roughness may occur due to "fouling" of the spark plug electrodes. This condition may occur on the ground or in the air and is usually the result of an excessively rich mixture setting which causes unburned carbon and lead deposits to collect on the spark plug electrodes. A pilot may alleviate this problem to some degree by always utilizing the recommended lean setting for the given condition.

G. Loss of Oil Pressure

1. During a cross-country flight you notice that the oil pressure is low, but the oil temperature is normal. What is the problem and what action should be taken?

A low oil pressure in flight could be the result of any one of several problems, the most common being that of insufficient oil. If the oil temperature continues to remain normal, a clogged oil pressure relief valve or an oil pressure gauge malfunction could be the culprit. In any case, a landing at the nearest airport is advisable to check for the cause of the trouble.

2. If a loss of oil pressure occurs accompanied by a rising oil temperature, what is indicated?

The oil required for cooling has been lost, and an engine failure is imminent. The throttle should be reduced, and a suitable landing area should be established as soon as possible. Use minimum power to reach the emergency landing area.

H. Smoke and Fire

1. What procedure should be followed if an engine fire develops on the ground during starting?

Continue to attempt an engine start as a start will cause flames and excess fuel to be sucked back through the carburetor.

a. *If the engine starts:*

    i. Increase the power to a higher rpm for a few moments; and then

    ii. Shut down the engine and inspect it.

b. *If the engine does not start:*

    i. Set the throttle to the "Full" position.

    ii. Set the mixture control to "Idle cutoff."

    iii. Continue to try an engine start in an attempt to put out the fire by vacuum.

c. *If the fire continues:*

    i. Turn the ignition switch to "Off."

    ii. Turn the master switch to "Off."

    iii. Set the fuel selector to "Off."

Evacuate the aircraft and obtain a fire extinguisher and/or fire personnel assistance.

2. What procedure should be followed if an engine fire develops in flight?

In the event of an engine fire in flight, the following procedure should be used:

a. Set the mixture control to "Idle cutoff."

b. Set the fuel selector valve to "Off."

c. Turn the master switch to "Off."

d. Set the cabin heat and air vents to "Off;" leave the overhead vents "On."

e. Establish an airspeed of 100 KIAS and increase the descent, if necessary, to find an airspeed that will provide for an incombustible mixture.

f. Execute a forced landing procedures checklist.

3. What procedure should be followed if an electrical fire develops inside the aircraft?

If an electrical fire is suspected (burning odor), the pilot should initially try to identify the possible source by checking all circuit breakers, avionics and instruments. If the problem is not detected and the odor or smoke continues, the following checklist should be completed:

a. Turn the master switch to "Off."
b. Set the avionics power switch to "Off."
c. Set all other switches to "Off" except the ignition switch.
d. Close all air/heat vents as well as any other air vents.
e. Use fire extinguisher.

4. What troubleshooting procedure should be followed in determining the cause of an electrical fire that is not readily apparent?

a. Turn the master switch to "Off."
b. Set all switches to "Off."
c. Check all circuit breakers for their status.

If the cause is still not readily apparent:

d. Turn the master switch to "On."
e. Set all switches to "On," one at a time, to isolate which circuit is tripping the breaker and/or causing the burning odor.
f. Once the faulty circuit is identified, disable it and ensure the fire is extinguished.
g. Open all air vents once the fire is extinguished.

5. What procedure should be followed if a cabin fire develops in flight?

Typically cabin fires are electrical in nature and identifying and disabling the faulty circuit is the first priority. However, careless smoking by passengers has also been a significant cause of cabin fires. The following checklist should be completed:

a. Turn the master switch to "Off."
b. Close all air/heat vents.
c. Use a fire extinguisher if available.
d. Land as soon as possible.

6. What procedure should be followed if a wing fire develops in flight?

If a wing fire develops in flight, the following checklist should be completed:

a. Set the navigation light switch to "Off."
b. Set the strobe light switch to "Off."
c. Set the pitot heat switch to "Off."

Initiate a sideslip maneuver to avoid flames from getting to the fuel tank and cabin area, then land as soon as possible.

I. Icing

1. What are the two main types of icing? (AC 00-6A)

Structural icing and induction icing.

2. Name four types of structural ice. (AC 00-6A)

Clear ice—Forms when large drops strike the aircraft surface and slowly freeze.

Rime ice—Small drops strike the aircraft and freeze rapidly.

Mixed ice—A combination of the above; usually supercooled water drops varying in size.

Frost—Ice crystal deposits formed by sublimation when the temperature at the surface is below the dew point, and the dew point is below freezing.

3. What is necessary for structural icing to occur? (AC 00-6A)

Visible moisture and below-freezing temperatures are necessary for structural icing to occur.

4. Which type of structural icing is more dangerous, rime or clear? (AC 00-6A)

Clear ice is hard, heavy, and tenacious. It is typically the most hazardous ice encountered. Clear ice forms when, after initial impact, the remaining liquid portion of the drop flows out over the aircraft surface, gradually freezing as a smooth sheet of solid ice. This type forms when drops are large, as in rain or in cumuliform clouds. Its removal by deicing equipment is especially difficult due to the fact that it forms as it flows away from the deicing equipment (inflatable boots, etc.).

5. What action is recommended if you inadvertently encounter icing conditions? (AC 00-6A)

You should change course and/or altitude; usually, climb to a higher altitude.

6. If icing has been inadvertently encountered, how would your landing approach procedure be different?

The following guidelines may be used when flying an airplane which has accumulated ice:

a. Maintain more power during the approach than normal.

b. Maintain a higher airspeed than normal.

c. Expect a higher stall speed than normal.

d. Expect a longer landing roll than normal.

e. A "no flaps" approach is recommended.

f. Maintain a consistently higher altitude than normal.

g. Avoid a missed approach (get it right the first time).

7. Which type of precipitation will produce the most hazardous icing conditions?

Freezing rain produces the most hazardous icing conditions.

8. Does the stall warning system have any protection from ice?

No, but some aircraft may be equipped with a heated stall warning system which consists of a vane, sensor unit and heating element on the leading edge of the wing. Usually this system is activated by the same switch that controls the pitot heat.

9. What causes "carburetor icing" and what are the first indications of its presence? (AC 61-23C)

The vaporization of fuel, combined with the expansion of air as it passes through the carburetor, causes a sudden cooling of the mixture. The temperature of the air passing through the carburetor may drop as much as 60°F within a fraction of a second. Water vapor is squeezed out by this cooling, and, if the temperature in the carburetor reaches 32°F or below, the moisture will be deposited as frost or ice inside the carburetor.

For airplanes with a fixed pitch propeller, the first indication of carburetor icing is a loss of rpm. For airplanes with controllable-pitch (constant speed) propellers, the first indication is usually a drop in manifold pressure.

10. What conditions are favorable for carburetor icing? (AC 61-23C)

On dry days, or when the temperature is well below freezing, the moisture in the air is not generally enough to cause trouble. But if the temperature is between 20°F and 70°F with visible moisture or high humidity, the pilot should be alert. Also during low or closed throttle settings, an engine is particularly susceptible to carburetor icing.

J. Pressurization

1. What is meant by decompression? (FAA-H-8083-3)

Decompression is the inability of the aircraft's pressurization system to maintain the designed "aircraft cabin" pressure. For example, an aircraft is flying at an altitude of 29,000 feet but the aircraft cabin is pressurized to an altitude equivalent to 8,000 feet. If decompression occurs, the cabin pressure may become equivalent to that of the aircraft's altitude of 29,000 feet. The rate at which this occurs determines the severity of decompression.

2. What are the two types of decompression? (FAA-H-8083-3)

Explosive decompression—Cabin pressure decreases faster than the lungs can decompress. Most authorities consider any kind of decompression which occurs in less than $\frac{1}{2}$ second as explosive and potentially dangerous. This type of decompression could only be caused by structural damage, material failure, or by a door "popping" open.

Rapid decompression—A change in cabin pressure where the lungs decompress faster than the cabin. There is no likelihood of lung damage in this case. This type could be caused by a failure or malfunction in the pressurization system itself, or through slow leaks in the pressurized area.

3. What are the dangers of decompression? (FAA-H-8083-3)

a. Hypoxia

b. At higher altitudes, being tossed or blown out of the airplane

c. Evolved gas decompression sickness (the bends)

d. Exposure to wind blast and extreme cold

K. Emergency Descent

1. When would an emergency descent procedure be necessary?

a. An in-flight fire

b. Depressurization in a pressurized airplane

c. A main door failure or window failure

d. Passenger or crewmember incapacitation

2. What procedure should be followed in establishing an emergency descent?

Generally the maneuver should be configured as recommended by the manufacturer. Except when prohibited by the manufacturer, the following procedure may be utilized:

a. Reduce power to idle;

b. Place propeller control in low pitch (high rpm)—acts as an aerodynamic brake;

c. As quickly as practical, extend landing gear/full flaps (maximum drag);

d. Establish a 30- to 45-degree bank for the purposes of clearing the area below.

L. Pitot Static System and Associated Flight Instruments

1. What instruments are affected when the pitot tube freezes? (AC 61-27C)

The airspeed indicator only—it acts like an altimeter; it will read higher as the aircraft climbs and lower as the aircraft descends. It reads lower than actual in level flight.

2. What instruments are affected when the static port freezes? (AC 61-27C)

Airspeed Indicator—Accurate at the altitude frozen as long as the static pressure in the indicator and the system equals outside pressure. If the aircraft descends, the airspeed indicator would read high (outside static pressure greater than that trapped). If the aircraft climbs, the airspeed indicator will read low.

Altimeter—Indicates the altitude at which the system was blocked.

Vertical speed—Indicates level flight.

3. Does the pitot system have any protection from ice?

Yes, the heated pitot system consists of the following components:

a. A pitot tube with a heating element,

b. A "pitot heat" rocker switch, and

c. A 10-amp push-to-reset circuit breaker.

4. What corrective action is needed if the pitot tube freezes? — If the static port freezes? (AC 61-27C)

Pitot tube—Turn pitot heat on.

Static port—Use alternate air if available, or break the face of a static instrument (either the VSI or A/S indicator).

5. What indications should you expect while using alternate air? (AC 61-27C)

In many unpressurized aircraft equipped with a pitot-static tube, an alternate source of static pressure is provided for emergency use. If the alternate source is vented inside the airplane, where static pressure is usually lower than outside static pressure, selection of the alternate static source may result in the following instrument indications:

Altimeter.................................... reads higher than normal
Airspeed indicator .... indicated airspeed reads greater than normal
Vertical velocity indicator .... momentarily shows a climb

M. Vacuum System and Associated Flight Instruments

1. **What instruments may be relied upon in the event of a complete vacuum system failure while operating in instrument meteorological conditions?**

   Turn and Slip/Turn Coordinator—bank information

   Magnetic Compass—bank information

   Airspeed—pitch information

   Altimeter—pitch information

   Vertical Speed Indicator—pitch information

2. **If the engine-driven vacuum pump were to fail, is there a backup system available?**

   Some general aviation aircraft may be equipped with a backup vacuum system. This system may be electrically driven or could be an engine-driven vacuum pump running in parallel to the primary pump.

N. Electrical

1. **What recommended procedure should be used in resetting a tripped circuit breaker?**

   Circuit breakers may be reset, but allow for a short cooling period to occur first (2 minutes). If, after resetting the circuit breaker, it pops out again, the circuit should be disabled and the circuit breaker left out.

2. Interpret the following ammeter indications.

a. Ammeter indicates a right deflection (positive).

After starting—The power from the battery used for starting is being replenished by the alternator. Or, if a full-scale charge is indicated for more than 1 minute, the starter is still engaged and a shutdown is indicated.

During flight—A faulty voltage regulator is causing the alternator to overcharge the battery.

b. Ammeter indicates a left deflection (negative).

After starting—Normal during start. Other times indicates the alternator is not functioning or an overload condition has occurred in the system. The battery is not receiving a charge.

During flight—The alternator is not functioning or an overload has occurred in the system. The battery is not receiving a charge.

3. What action should be taken if the ammeter indicates a continuous discharge (left needle) while in flight?

The alternator has quit producing a charge, so the master switch and the alternator circuit breaker should be checked and reset if necessary. If this does not correct the problem, the following should be accomplished:

a. The alternator should be turned off; pull the circuit breaker (field circuit will continue to draw power from the battery).

b. All electrical equipment not essential to flight should be turned off (the battery is now the only source of electrical power).

c. The flight should be terminated and a landing made as soon as possible.

4. What action should be taken if the ammeter indicates a continuous charge (right needle) while in flight (more than two needle widths)?

If a continuous excessive rate of charge were allowed for any extended period of time, the battery would overheat and evaporate the electrolyte at an excessive rate. A possible explosion of the battery could result. Also, electronic components in the electrical system would be adversely affected by higher than normal voltage.

Protection is provided by an overvoltage sensor which will shut the alternator down if an excessive voltage is detected. If this should occur the following should be done:

a. The alternator should be turned off; pull the circuit breaker (the field circuit will continue to draw power from the battery).

b. All electrical equipment not essential to flight should be turned off (the battery is now the only source of electrical power).

c. The flight should be terminated and a landing made as soon as possible.

5. If the low-voltage warning light illuminates, what has occurred?

The alternator has been shut down; the airplane is equipped with a combination alternator/regulator high-low voltage control unit and a red warning light labeled Low Voltage. In the event of an overvoltage condition, the alternator control unit will shut down the alternator. The battery will then supply system current. The low-voltage warning light will illuminate when the system voltage drops below normal. Follow the procedure for continuous discharge inflight.

O. Landing Gear

1. If a positive gear down indication is not received, what action is recommended first?

Several preliminary checks can be made prior to utilizing the emergency extension procedure:

a. Check that the master switch is set to "On."

b. Check that the "Landing Gear" and "Gear Pump" circuit breakers are in.

c. Check both "Landing Gear" position indicators by using the "Press-To-Test" feature and by rotating the dimming shutter.

d. If a bulb has burned out, you can use the other operating bulb as a temporary replacement.

2. What recommended procedure should be utilized if the landing gear fails to retract after takeoff?

If the landing gear fails to retract normally, the following checklist should be completed.

a. Check that the master switch is set to "On."

b. Check that the landing gear lever is in the full up position.

c. Check that the gear pump and landing gear circuit breakers are "In."

d. Check the gear up light.

e. Recycle the landing gear lever.

f. Check for proper gear motor operation by examining the ammeter and listening for noise.

Note: If you still hear a gear motor noise after 1 minute, pull out the gear pump circuit breaker to avoid overheating the motor. You can reinstall the circuit breaker when needed for landing.

3. How is the emergency gear extension system operated?

There is a hand-operated pump, located between the front seats, which may be used for manual extension of the landing gear in the event of a hydraulic failure.

4. What is the recommended procedure if the landing gear will not extend normally?

If the landing gear fails to extend normally, the following checklist should be completed.

a. Check that the master switch is "On."

b. Check that the landing gear lever is "Down."

c. Check that the gear pump and landing gear circuit breakers are "In."

d. Extend the handle and pump the emergency hand pump until heavy resistance is encountered (about 30–40 times).

e. Check that the gear down light is "On."

f. Secure the pump handle.

5. What procedure should be followed if a pilot does not receive a positive indication of the gear being down and locked?

Attempt to extend the gear manually. If this action is unsuccessful, plan for a gear-up landing. The following checklist should be completed:

a. Complete the "before landing" checklist.

b. Establish a normal approach configuration with full flaps.

c. Check that the gear pump and landing gear circuit breakers are "In."

d. Initiate a tail low landing.

e. Use a minimum amount of braking.

f. Taxi slowly.

g. Shut down the engine and then inspect the gear.

6. What is the recommended procedure in dealing with a flat main landing gear tire?

a. Establish a normal approach configuration with full flaps.

b. Touchdown with the good tire first on that side of the runway and keep the aircraft off of the flat tire for as long as possible.

c. Use braking on the good wheel as required to maintain directional control.

7. What is the recommended procedure to follow if the nose gear is unsafe or the tire is flat?

a. Complete the before landing checklist.

b. Shift weight to the rear by moving passengers and/or baggage to the rear.

c. Set the flaps to the 30° position.

d. Unlatch all doors.

e. After committing to a landing, set both the avionics and master switch to the "Off" position.

f. Initiate touchdown in a slightly tail low configuration.

Continued

g. Set the mixture control to the "Idle-Cutoff" position.

h. Set the ignition switch to the "Off" position.

i. Set the fuel selector to the "Off" position.

j. Hold the nose off as long as possible.

k. After landing, evacuate the aircraft as soon as possible.

8. Why should taxiing on a slush, snow, or ice covered taxiway in a retractable gear airplane be avoided? (AC 91-6A)

Problems may occur with the retraction or extension of the landing gear due to water, slush, or snow freezing on various gear system components. On the climbout, recycle the landing gear two or three times to prevent gear components from freezing.

P. Wing Flaps (Asymmetrical Position)

1. What is an "asymmetrical" flap emergency?

This emergency situation occurs when, for one of the following reasons, the flap position on one wing is different from the flap position on the other wing. When this occurs, a strong rolling tendency will result. The following are possible factors contributing to this situation:

a. Mechanical failure of the flap extension mechanism

b. Faulty maintenance performed

c. Maximum flap extension speed exceeded

2. What procedure should be followed in an asymmetrical flap emergency?

The rolling tendency can be counteracted with immediate opposite aileron followed by a timely retraction of the extended flap. This situation can be avoided by limiting flap extension and retraction to smaller increments allowing detection and correction before the rolling tendency has fully developed.

Q. Inoperative Elevator

What procedure should be followed if loss of elevator control occurs?

a. Extend the landing gear.

b. Lower flaps by 10°.

c. Set trim for level flight.

d. Using throttle and elevator trim control, establish an airspeed of 70 knots.

Do not change the established trim setting. Maintain control of the glide angle by adjusting power. At the landing flare, the elevator trim should be adjusted to full noseup and the power reduced. At the moment of touchdown, close the throttle.

R. Inadvertent Door Opening

1. What procedure should be followed if a cabin door accidentally opens in flight?

a. Maintain control of the aircraft.

b. Establish a straight and level flight configuration.

c. Trim the aircraft.

d. Reduce the airspeed.

e. Make an attempt to close the door.

f. Close the cabin air vents to reduce cabin pressure.

g. If the door won't close, a controlled slip may help.

2. What procedure should be followed if a baggage door opens in flight?

Baggage compartments tend to be located in the aft section of the airplane and under no circumstances should the pilot allow anyone to attempt to close the doors while in flight. By design, the baggage compartment door will tend to remain closed during flight due to airflow pressure.

S. Emergency Equipment and Survival Gear

1. What two factors should be considered when considering the type of survival equipment to carry for a flight over an uninhabited area? (AIM 6-2-7)

    a. The type of climate

    b. The type of terrain

2. What additional equipment is required if an aircraft is operated for hire over water and beyond power-off gliding distance from shore? (14 CFR 91.205)

If an aircraft is operated for hire over water and beyond power-off gliding distance from shore, approved flotation gear readily available to each occupant and at least one pyrotechnic signaling device.

3. What do you have in the aircraft that can be used to aid in survival?

    a. The compass will keep you going in one direction.

    b. Gasoline will help make a fire.

    c. Oil can be used for smoke signals.

    d. Seat upholstery may be used to wrap around feet or hands.

    e. Wiring may be used for tie strings.

    f. The battery may be used to ignite fuel.

Additional Study Questions

1. In the event of an emergency requiring a forced landing, what information should be included in the "emergency" briefing to the passengers? (POH)

2. If an engine failure occurs immediately after takeoff, why is it inadvisable to turn back to the departure runway for landing? (FAA-H-8083-3)

3. During an emergency landing you realize that you have misjudged the glidepath and will "overshoot" the forced landing area. What procedures should be used? What procedures should be used for an "undershoot"? (FAA-H-8083-3)

4. What effect does the wind have on an emergency approach and landing procedure? (FAA-H-8083-3)

5. If the landing gear will not retract normally, can you retract the gear manually? How? (POH)

6. You have just extended the landing gear manually due to a hydraulic system failure (suspected loss of hydraulic fluid). Should you expect to also have braking problems on the landing rollout? (POH)

7. If the alternator has failed and the battery is the only source of electrical power, approximately how much time remains before a complete electrical system failure occurs? What factors will affect the time remaining? (POH)

8. What flight instrument indications would you see if both the pitot static tube and static port became blocked for some reason? (AC 61-27C)

9. What procedures should you utilize if the trim control on your aircraft has become inoperative, resulting in excessive control pressures inflight? (POH)

10. Are fuel injected engines subject to induction system icing? (AC 61-23C)

Determining Performance and Limitations

5

A. Advanced Aerodynamics

1. What is the "chord line" of a wing? (AC 61-23C)

A reference line often used in discussing the airfoil is the "chord line," a straight line drawn through the profile connecting the extremities of the leading and trailing edges.

2. What is the "camber" of a wing? (AC 61-23C)

Camber is simply the curvature of the wing. The distance from the chord line to the upper and lower surfaces of the wing denotes the magnitude of the upper and lower camber at any point.

3. What is "angle of attack"? (AC 61-23C)

Angle of attack is the angle between the chord line of the wing and the relative wind.

4. Will the critical angle of attack at which an airplane stalls change with weight, bank angle, etc.? (AC 61-23C)

For any given airplane, the stalling or critical angle of attack remains constant regardless of weight, dynamic pressure, bank angle, or pitch attitude. These factors will affect the speed at which the stall occurs, but not the angle.

5. What is the critical angle of attack in degrees for most aircraft? (AC 61-23C)

18° to 20° is the critical angle of attack for most wing designs.

6. What is "angle of incidence"? (AC 61-23C)

The angle of incidence is the angle at which the wing is attached relative to the longitudinal axis of the airplane.

7. What is the "center of pressure"? (AC 61-23C)

Sometimes called "center of lift," it is the point along the chord line on an airfoil where lift is concentrated. In general, at high angles of attack the center of pressure moves forward, while at low angles of attack the center of pressure moves aft. The airplane's aerodynamic balance and controllability are governed by changes in the center of pressure.

8. Discuss the relationship of center of gravity to center of pressure. (AC 61-23C)

The location of the center of gravity is determined by the general design of each airplane. The designers determine how far the center of pressure will travel. They then fix the center of gravity forward of the center of pressure for the corresponding flight speed in order to provide an adequate restoring moment to retain flight equilibrium.

9. What are several factors which will affect both lift and drag? (AC 61-23C)

Wing area—Lift and drag acting on a wing are roughly proportional to the wing area. A pilot can change wing area by using certain types of flaps (i.e., Fowler flaps).

Shape of the airfoil—As the upper curvature of an airfoil is increased (up to a certain point) the lift produced increases. Lowering an aileron or flap device can accomplish this. Also, ice or frost on a wing can disturb normal airflow, changing its camber and disrupting its lifting capability.

Angle of attack—As angle of attack is increased, both lift and drag are increased, up to a certain point.

Velocity of the air—An increase in velocity of air passing over the wing increases lift and drag.

Air density—Lift and drag vary directly with the density of the air. As air density increases, lift and drag increase. As air density decreases, lift and drag decrease. Air density is affected by several factors: pressure, temperature, and humidity.

10. What is a wing planform? (AC 61-23C)

The wing planform is the shape of the wing as viewed from directly above. It deals with airflow in three dimensions, and is very important to understanding wing performance and airplane flight characteristics. Aspect ratio, taper ratio, and sweepback are factors in planform design that are very important to the overall aerodynamic characteristic of a wing.

11. What are the four common wing designs? (AC 61-23C)

Elliptical—Provides a minimum of induced drag; stall characteristics inferior to a rectangular wing; difficult to construct.

Rectangular—Good stall characteristics, stalling at the wing root first; provides adequate aileron effectiveness; usually quite stable; favored design for low-cost, low-speed airplanes.

Tapered—Desirable from a standpoint of weight and stiffness; good stall characteristics; comparable in efficiency to the elliptical wing; cheaper to build than an elliptical wing.

Sweptback—Generally for higher speed aircraft; very different stall characteristics; requires very precise flying techniques.

12. What is aspect ratio? (AC 61-23C)

Aspect ratio is the ratio of wing span to wing chord. This ratio is the primary factor in determining the three-dimensional characteristics of the ordinary wing and its lift/drag ratio.

13. What is "sweepback"? (AC 61-23C)

Sweepback is the rearward slant of a wing, horizontal tail, or other airfoil surface. It is utilized mainly in high-speed aircraft design. At speeds above 450 knots, a significant increase in drag occurs due to the compression of air at the leading edge. Sweepback delays this sharp increase in drag to a higher airspeed.

14. What is meant by "tapering" a wing? (AC 61-23C)

"Tapering" a wing is a means of changing a wing planform by decreasing the length of the chord from the root to the tip of the wing. In general, tapering will cause a decrease in drag (most effective at high speeds) and an increase in lift. There is also a structural benefit due to savings in the weight of the wing.

15. In designing a wing, why is it important that the wing root stall first? (AC 61-23C)

Most straight aircraft wings are designed so that the wing root stalls before the outer section stalls. This enables the ailerons to remain somewhat effective for lateral control of roll during the stall and also enhances aircraft stability.

16. What are several methods a manufacturer may employ to cause the wing to stall at the root first?

a. The wing is designed so that the outer section has less angle of incidence than the inner section or wing root. Sometimes this is referred to as "washout" or "twist."

b. The outer section of the airfoil is designed to produce slightly more lift than the inner section.

c. Stall strips may be attached to the leading edge of a wing near the wing root. At higher angles of attack, these stall strips will cause a general disruption of airflow resulting in the wing root stalling first.

17. Name four basic types of flaps. (AC 61-23C)

Plain flap—Most common flap system; when extended it increases the camber of the wing which results in an increase in both lift and drag.

Split flap—Similar to the plain flap; the main difference is the split flap produces more drag which enables a steeper approach without an increase in airspeed.

Slotted flap—A slotted flap will produce proportionally more lift than drag. Its design allows high pressure air below the wing to be directed through a slot to flow over the upper surface of the flap, delaying the airflow separation at higher angles of attack. This design lowers the stall speed significantly.

Fowler flap—Very efficient design. Moves backward on first part of extension increasing lift with little drag; also utilizes a slotted design resulting in lower stall speeds.

18. When lowering the flaps, why do some aircraft encounter a pitch change? (AC 61-23C)

Use of flaps alters the lift distribution and consequently causes the center of pressure to move aft. Because of this movement, a nose-down pitching moment may be experienced. Low-wing aircraft are more subject to this characteristic than high-wing airplanes. With flap extension on a high-wing airplane, the center of pressure also moves aft but the airflow from the wing is directed downward which pushes down on the horizontal stabilizer. This results in a counteracting nose-up pitching moment.

19. What is parasite drag? (AC 61-23C)

Parasite drag applies to the entire airplane and is composed of forces caused by protuberances extending into the airstream which do not contribute to lift. Items such as radio antennas, struts, fittings, landing gear, etc. produce parasite drag. Parasite drag is greatest at high airspeeds and is proportional to the square of the airspeed. If the airspeed were doubled, the parasite drag would be quadrupled.

20. What is induced drag? (AC 61-23C)

Induced drag results from the production of lift. The amount of induced drag varies inversely with the airspeed. The lower the airspeed the greater the angle of attack required to produce lift equal to the airplane's weight and consequently, the greater will be the induced drag.

21. As airplane speed increases, how much of an increase in drag will there be? (AC 61-23C)

As the speed increases from stalling speed to top speed, the induced drag decreases and parasite drag increases. If an airplane in steady level flight condition accelerates from 100 knots to 200 knots, parasite drag becomes four times as great and induced drag is only one-fourth its original value.

22. What is lift/drag ratio? (AC 61-23C)

The ratio of the total amount of lift being produced to the total amount of drag being produced at a given airspeed. The maximum L/D ratio determines the airspeed at which the most lift is produced for the least amount of drag. Also, this speed will produce the best power-off gliding distance (best glide speed), and the maximum range available for the aircraft.

23. What are wing-tip vortices? (Pilot/Controller Glossary)

Circular patterns of air created by the movement of an airfoil through the air when generating lift. As an airfoil moves through the atmosphere in sustained flight, an area of low pressure is created above it. The air flowing from the high-pressure area to the low-pressure area around and about the tips of the airfoil tends to roll up into two rapidly counter-rotating vortices, cylindrical in shape. These vortices are the most predominant parts of aircraft wake turbulence.

24. What factors determine the strength of wing-tip vortices? (AIM 7-3-3)

The strength of the vortex is governed by the weight, speed, and shape of the wing of the generating aircraft. The greatest vortex strength occurs when the generating aircraft is HEAVY, CLEAN, and SLOW.

25. What is "ground effect"? (AC 61-23C)

Ground effect occurs when an aircraft, operating approximately one wing span above the surface, experiences a reduction in induced drag and a resultant increase in the efficiency of the wing. This is due to the interference of the ground with the airflow patterns about the airplane, or more specifically the wing's upwash, downwash, and wing-tip vortices. In simple terms, ground effect is the cushioning or pushing effect of the air as it is compressed by an airplane flying close to the ground. Its effects are of greatest concern during landings and takeoffs.

26. What major problems can be caused by ground effect?

During landing—At a height of approximately one-tenth of a wing span above the surface, drag may be 40 percent less than when the airplane is operating out of ground effect. Therefore, any excess speed during the landing phase may result in a significant float distance. In such cases, if care is not exercised by the pilot, he/she may run out of runway and options at the same time.

During takeoff—Due to the reduced drag in ground effect, the aircraft may seem capable of takeoff well below the recommended speed. However, as the airplane rises out of ground effect with a deficiency of speed, the greater induced drag may result in very marginal climb performance, or the inability of the airplane to fly at all. In extreme conditions such as high temperature, high gross weight, and high density altitude, the airplane may become airborne initially with a deficiency of speed and then settle back to the runway.

27. Define the three aircraft axes. (AC 61-23C)

Longitudinal axis—Extends lengthwise through the fuselage from the nose to the tail. Rolling moments occur on this axis.

Vertical axis—Passes vertically through the center of gravity. Yawing moments occur on this axis.

Lateral axis—Extends crosswise from wingtip to wingtip. Pitching moments occur on this axis.

28. What are the two basic types of stability? (AC 61-23C)

a. Static stability
b. Dynamic stability

29. What is static stability? (AC 61-23C)

Static stability is the initial tendency of an airplane to return to its original attitude after it has been displaced. It may be one of the following types:

a. *Positive static stability*—The tendency of an aircraft to return to its original state of equilibrium after being disturbed. Most aircraft possess this design characteristic.

b. *Negative static stability*—The tendency of an aircraft to continue away from its original state of equilibrium.

c. *Neutral static stability*—The tendency of an aircraft to remain in a new condition after its equilibrium has been disturbed.

30. What is dynamic stability? (FAA-H-8083-3)

Dynamic stability is the overall tendency that an airplane displays after its equilibrium has been disturbed. It may take the form of:

a. *Longitudinal phugoid oscillations*—The angle of attack will remain constant while the airspeed increases or decreases.

b. *Short period longitudinal oscillations*—Usually impossible to control by the pilot because of their short duration. Changes in angle of attack occur with no change in airspeed. These types of oscillations are very significant because they can cause structural failure if not damped immediately—which is an effect of dynamic stability required by the regulations.

31. What is longitudinal stability? (AC 61-23C)

Longitudinal stability is the quality which makes an airplane stable about its lateral axis. It involves the pitching motion as the airplane's nose moves up and down in flight. A longitudinally unstable airplane has a tendency to dive or climb progressively into a very steep dive or climb, or even a stall. An airplane with longitudinal instability will be difficult and sometimes dangerous to fly.

32. What variable factor affects longitudinal stability for a given aircraft? (AC 61-23C)

The location of the center of gravity will have a significant effect on stability. In general, an airplane will be more stable with a forward CG and less stable with an aft CG.

33. What is "lateral stability"? (AC 61-23C)

Stability about the airplane's longitudinal axis, which extends from the nose to the tail, is called lateral stability. This helps to stabilize the lateral or rolling effect when one wing gets lower than the other.

34. What is "dihedral"? (AC 61-23C)

The most common procedure for producing lateral stability is to build the wings with a dihedral angle varying from 1° to 3°. In other words, the wings on either side of the airplane join the fuselage to form a slight "V" or angle called "dihedral," and this is measured by the angle made by each wing above a line parallel to the lateral axis.

35. What is directional stability? (AC 61-23C)

Stability about the airplane's vertical axis (the sideways moment) is called yawing or directional stability.

36. How is directional stability achieved in airplane design? (AC 61-23C)

Yawing or directional stability is the more easily achieved stability in airplane design. The area of the vertical stabilizer and the sides of the fuselage aft of the center of gravity are the prime contributors.

37. What determines radius of turn? (AC 61-23C)

Angle of bank and speed of the aircraft. If angle of bank increases or airspeed decreases, the radius of turn decreases.

38. What causes "adverse yaw"? (AC 61-23C)

When turning an airplane to the left, for example, the downward deflected aileron on the right produces more lift on the right wing. Since the downward deflected right aileron produces more lift, it also produces more drag, while the opposite left aileron has less lift and less drag. This added drag attempts to pull or veer the airplane's nose in the direction of the raised wing (right); that is, it tries to turn the airplane in the direction opposite to that desired. This undesired veering is referred to as adverse yaw.

39. How is adverse yaw controlled in aircraft design? (AC 61-23C)

An aileron may be designed to have more upward movement than downward movement. This will result in less drag on the down aileron and an increase in drag on the up aileron.

40. What are several factors which would result in a higher stalling speed? (AC 61-23C)

a. Excessive weight

b. A forward CG

c. No flaps

d. Frost, snow or ice

e. Turbulence

f. An increase in angle of bank

g. An increase in load factor

h. An uncoordinated turn

41. Does an increase in altitude have an effect on the indicated airspeed at which an airplane stalls? (AC 61-23C)

An increase in altitude has no effect on the indicated airspeed at which an airplane stalls. The indicated stalling speed remains the same.

42. Does an increase in altitude affect the true airspeed at which an airplane stalls? (AC 61-23C)

Since true airspeed normally increases as altitude increases (for a given indicated airspeed), then the true airspeed at which an airplane stalls generally increases with an increase in altitude. Under nonstandard conditions (temperature lower than standard) there is an additional increase in true airspeed above the indicated airspeed.

43. What is "load factor"? (FAA-H-8083-3)

Load factor is the ratio of the total airload acting on the airplane to the gross weight of the airplane. Any force applied to an airplane to deflect its flight from a straight line produces a stress on its structure; the amount of this force is termed "load factor." For example, a load factor of 3 Gs means that the total load on an airplane's structure is three times its gross weight.

44. For what two reasons is load factor important to pilots? (FAA-H-8083-3)

Load factor is important

a. because of the obviously dangerous overload that is possible for a pilot to impose on the aircraft structure.

b. because an increased load factor increases the stalling speed and makes stalls possible at seemingly safe flight speeds.

45. What situations may result in the load factor reaching the maximum or being exceeded? (FAA-H-8083-3)

Turns—The load factor increases at a terrific rate after a bank has reached 45 or 50°. The load factor in a 60°-bank turn is 2 Gs. The load factor in a 80°-bank turn is 5.76 Gs. The wing must produce lift equal to these load factors if altitude is to be maintained.

Turbulence—Severe vertical gusts can cause a sudden increase in angle of attack, resulting in large loads which are resisted by the inertia of the airplane.

Continued

Speed—The amount of excess load that can be imposed upon the wing depends on how fast the airplane is flying. At speeds below maneuvering speed, the airplane will stall before the load factor can become excessive. At speeds above maneuvering speed, the limit load factor for which an airplane is stressed can be exceeded by abrupt or excessive application of the controls or by strong turbulence.

46. Does the maneuvering speed change with weight? (AC 61-23C)

Yes; when an aircraft is operated at a weight less than gross weight, the maneuvering speed will be less. A general rule of thumb is "a 2 percent change in weight will result in a 1 percent change in maneuvering speed."

B. Aircraft Performance

1. The performance data given for your aircraft is based on what conditions?

It is based on a standard atmosphere (15°C and 29.92" Hg).

2. What are some of the main elements of aircraft performance? (FAA-H-8083-3)

a. Takeoff and landing distance

b. Rate-of-climb

c. Ceiling

d. Payload

e. Range

f. Speed

g. Fuel economy

3. What factors affect the performance of an aircraft during takeoffs and landings? (FAA-H-8083-3)

a. Air density (density altitude)

b. Surface wind

c. Runway surface

d. Upslope or downslope of runway

e. Weight

4. How does surface wind affect takeoff and landing performance? (AC 61-23C)

For takeoffs the effect of wind is great. A headwind will allow an airplane to reach liftoff speed at a lower ground speed, reducing the takeoff distance significantly. A tailwind will require the airplane to achieve a greater ground speed to attain the liftoff speed, increasing the effective takeoff distance.

Wind also has a great effect on landings. A headwind will shorten the landing distance by decreasing the ground speed at which the airplane touches down. A tailwind will significantly increase the landing distance due to an increase in ground speed of the aircraft as it approaches and touches down.

5. How does weight affect takeoff and landing performance? (AC 61-23C)

Increased gross weight can be considered to produce these effects:

a. Higher liftoff and landing speed required;

b. Greater mass to accelerate or decelerate (slow acceleration/ deceleration);

c. Increased retarding force (drag and ground friction); and

d. Longer takeoff and ground roll.

The effect of gross weight on landing distance is that the airplane will require a greater speed to support the airplane at the landing angle of attack and lift coefficient resulting in an increased landing distance.

6. How does air density affect takeoff and landing performance? (AC 61-23C)

An increase in density altitude (decrease in air density) can produce these effects:

a. Higher liftoff and landing speed required (same indicated airspeed, higher true airspeed);

b. Decreased thrust and reduced acceleration;

c. Longer takeoff and landing roll; and

d. Decreased climb rate.

7. Define the term "density altitude." (FAA-H-8083-3)

Density altitude is pressure altitude corrected for nonstandard temperature. It is the measurement of air density in terms of an altitude in a standard atmosphere.

8. How does air density affect aircraft performance? (FAA-H-8083-3)

The density of the air has a direct effect on

a. Lift produced by the wings;

b. Power output of the engine;

c. Propeller efficiency; and

d. Both induced and parasite drag.

9. How do temperature, altitude, humidity and barometric pressure affect density altitude? (FAA-H-8083-3)

Density altitude will *increase* (low air density) when one or more of the following occurs:

a. High air temperature

b. High altitude

c. High humidity

d. Low barometric pressure

Density altitude will *decrease* (high air density) when one or more of the following occurs:

a. Low air temperature

b. Low altitude

c. Low humidity

d. High barometric pressure

10. What effect does landing at high-elevation airports have on ground speed with comparable conditions relative to temperature, wind and airplane weight?

Even though using the same indicated airspeed that is appropriate for sea level operations, the true airspeed is faster, resulting in a faster ground speed (with a given wind condition) throughout the approach, touchdown, and landing roll. This results in a greater distance to clear obstacles during the approach, a longer ground roll, and consequently the need for a longer runway. All of these factors should be taken into consideration when landing at high-elevation fields, particularly if the field is short.

11. What performance is characteristic of flight at maximum L/D in a propeller-driven airplane?

The maximum range condition and maximum glide ratio will be obtained when operating at L/D_{MAX} in a reciprocating-engine aircraft.

12. Define the terms "maximum range" and "maximum endurance." (AC 61-23C)

Maximum range is the maximum distance an airplane can fly for a given fuel supply, and is obtained at the maximum lift/drag ratio (L/D_{MAX}). It is important to note that for a given airplane configuration, the maximum lift/drag ratio occurs at a particular angle of attack and lift coefficient, and is unaffected by weight or altitude.

Maximum endurance is the maximum amount of time an airplane can fly for a given fuel supply and is obtained at the point of minimum power required since this would require the lowest fuel flow to keep the airplane in steady, level flight.

13. Why does the manufacturer provide various manifold pressure/prop settings for a given power output?

The various power MAP/rpm combinations are provided so the pilot has a choice between operating the aircraft at best efficiency (minimum fuel flow) or operating at best power/speed condition. An aircraft engine operated at higher rpms will produce more friction and as a result use more fuel. On the other hand, an aircraft operating at higher and higher altitudes will not be able to continue to produce the same constant power output due to a drop in manifold pressure. The only way to compensate for this is by operating the engine at a higher rpm.

14. What does the term 75% brake horsepower mean?

Brake horsepower (BHP) is the power delivered at the propeller shaft (main drive or main output) of an aircraft engine. 75% BHP means you are delivering 75 percent of the normally rated power or maximum continuous power available at sea level on a standard day to the propeller shaft.

15. Explain how 75% BHP can be obtained from your engine.

Set the throttle (manifold pressure) and propeller (rpm) to the recommended values found in the Cruise Performance Chart of your Pilot's Operating Handbook.

16. When would a pilot lean a normally aspirated direct-drive engine? (FAA P-8740-13)

a. Lean anytime the power setting is 75 percent or less at any altitude.

b. At high-altitude airports, lean for taxi, takeoff, traffic pattern entry and landing.

c. When the density altitude is high (Hot, High, Humid).

d. For landings at airports below 5,000 feet density altitude, adjust the mixture for descent, but only as required.

e. In any event, always consult your Pilot's Operating Handbook for the proper leaning procedures.

17. What are the different methods available for leaning aircraft engines? (FAA P-8740-13)

Tachometer Method—For best economy operation, the mixture is first leaned from full rich to maximum power (peak rpm), then the leaning process is slowly continued until the engine runs rough. Then enrich the mixture sufficiently to obtain a smoothly firing engine.

Fuel Flowmeter Method—Aircraft equipped with fuel flowmeters require that you lean the mixture to the published (POH) or marked fuel flow to achieve the correct mixture.

Exhaust Gas Temperature Method—Lean the mixture slowly to establish peak EGT then enrich the mixture by 50 degrees rich (cooler) of peak EGT. This will provide the recommended lean condition for the established power setting.

18. Define the following airplane performance speeds. (FAA-H-8083-3)

V_{S0}—Stall speed in the landing configuration; the calibrated power-off stalling speed or the minimum steady flight speed at which the airplane is controllable in the landing configuration.

V_S—Stall speed clean or in a specified configuration; the calibrated power-off stalling speed or the minimum steady flight speed at which the airplane is controllable in a specified configuration.

V_Y—Best rate-of-climb speed; the calibrated airspeed at which the airplane will obtain the maximum increase in altitude per unit of time. This best rate-of-climb speed normally decreases slightly with altitude.

V_X—Best angle-of-climb speed; the calibrated airspeed at which the airplane will obtain the highest altitude in a given horizontal distance. This best angle-of-climb speed normally increases with altitude.

V_{LE}—Maximum landing gear extension speed; the maximum calibrated airspeed at which the airplane can be safely flown with the landing gear extended. This is a problem involving stability and controllability.

Continued

V_{LO}—Maximum landing gear operating speed; the maximum calibrated airspeed at which the landing gear can be safely extended or retracted. This is a problem involving the airloads imposed on the operating mechanism during extension or retraction of the gear.

V_{FE}—Maximum flap extension speed; the highest calibrated airspeed permissible with the wing flaps in a prescribed extended position. This is a problem involving the airloads imposed on the structure of the flaps.

V_A—Maneuvering speed; the calibrated design maneuvering airspeed. This is the maximum speed at which the limit load can be imposed (either by gusts or full deflection of the control surfaces) without causing structural damage.

V_{NO}—Normal operating speed; the maximum calibrated airspeed for normal operation or the maximum structural cruise speed. This is the speed at which exceeding the limit load factor may cause permanent deformation of the airplane structure.

V_{NE}—Never exceed speed; the calibrated airspeed which should *never* be exceeded. If flight is attempted above this speed, structural damage or structural failure may result.

19. **The following questions are designed to provide the pilot with a general review of the basic performance information he/she should know about his/her specific airplane before taking a flight check or review.**

    a. **What is the normal climb-out speed?**
    b. **What is the best rate-of-climb speed?**
    c. **What is the best angle-of-climb speed?**
    d. **What is the maximum flap extension speed?**
    e. **What is the maximum gear extension speed?**
    f. **What is the maximum gear retraction speed?**
    g. **What is the stall speed in the normal landing configuration?**
    h. **What is the stall speed in the clean configuration?**
    i. **What is the normal approach-to-land speed?**
    j. **What is the maneuvering speed?**
    k. **What is the red-line speed?**
    l. **What speed will give you the best glide ratio?**
    m. **What is the maximum window-open speed?**

  n. **What is the maximum allowable crosswind compo-
        nent for the aircraft?**
  o. **What takeoff distance is required if a takeoff were
        made from a sea level pressure altitude?**

C. Aircraft Performance Charts

The following are typical questions concerning aircraft performance
charts. Refer to your Pilot's Operating Handbook for the answers.

1. Takeoff and Landing Distance Charts

  a. **Given the following conditions:**
        Pressure Altitude = 4,000 feet
        Temperature = 40°F
        Runway = Hard Surfaced
        Weight = Maximum Takeoff Weight
        Wind = 10 Knot Headwind
        What is the distance for a normal takeoff ground roll?
        What is the distance to clear a 50-foot obstacle?
  b. **Given the following conditions:**
        Pressure altitude = 2,000 feet
        Temperature = 25°C
        Runway = Hard surfaced
        Weight = Maximum Landing Weight
        Wind = Calm
        What is the normal landing distance?
        **What is the minimum landing distance over a 50-foot
        obstacle?**

2. Time, Fuel, and Distance-to-Climb Chart

  Given the following conditions:
        Climb = 2,000 feet to 7,000 feet Pressure altitude
        Temperature = Standard

Continued

Airspeed = Best Rate-of-Climb

Wind speed = Calm

How much time is required for the climb?

How much fuel is required for the climb?

What distance will be covered during the climb?

3. **Maximum Rate-of-Climb Chart**

Given the following conditions:

Pressure Altitude = 4,000 feet

Outside Air Temperature = 0°C

What is the best rate-of-climb airspeed?

What is the best rate-of-climb, in feet per minute?

4. **Cruise Performance Chart**

a. Given the following conditions:

Pressure Altitude = 6,000 feet

Engine rpm = 2,400

MAP = 22 inches

Temperature = Standard

What will the true airspeed be?

What will the fuel consumption rate be?

What will the percent brake horsepower be?

b. Given the following conditions:

Density Altitude at cruising altitude = 4,000 feet

Temperature = Standard

What percentage of brake horsepower will be required to produce the maximum true airspeed?

What will the required power setting be?

5. Maximum Range/Endurance Chart

a. Given the following conditions:

Density Altitude = 8,000 feet

Temperature = Standard

Weight = Maximum Gross

Wind = Calm

Power = 75%

BHP Fuel = Full Tanks (Standard) 45 minute reserve

What true airspeed can be expected?

What maximum range will be achieved?

b. Using the data from the previous question, how long can the aircraft fly in hours?

6. Miscellaneous

a. Determine the approximate CAS you should use to obtain 180 knots TAS with a pressure altitude of 8,000 feet and a temperature of +4°C.

158 knots.

b. At speeds below 200 knots (where compressibility is not a factor), how is true airspeed computed?

True airspeed can be found by correcting calibrated airspeed for pressure altitude and temperature.

c. Compute the density altitude for the following conditions:

Temperature   = 20°C

Field Elevation   = 4,000 feet

Altimeter setting = 29.98

Continued

d. Compute the standard temperature at 9,000 feet.

The standard temperature at sea level is 15°C. The average lapse rate is 2° per 1,000 feet. Compute standard temperature by multiplying the altitude by 2 and then subtracting that number from 15. Based on this information,

$$15° - (2° \times 9 = 18°) = -3°C$$

The standard temperature at 9,000 feet is -3°C.

e. A descent is planned from 8,500 feet MSL when 20 NM from your destination airport. If ground speed is 150 knots and you desire to be at 4,500 feet MSL when over the airport, what should the rate of descent be?

i.   Change in altitude = 4,000 feet

ii. Calculate time to go 20 NM at 150 knots (8 minutes)

iii. 4,000 feet ÷ 8 minutes = 500 FPM

f. A descent is planned from 11,500 feet MSL to arrive at 7,000 feet MSL, 5 SM from a VORTAC. With a ground speed of 160 mph and a rate of descent of 600 FPM, at what distance from the VORTAC should the descent be started?

i.   Change in altitude = 4,500 feet

ii. Rate of Descent = 600 FPM

iii. Time to descend = 4,500 ÷ 600 = 7.5 minutes

iv. Ground speed in miles per minute =
     160 ÷ 60 = 2.67 MPM

v.   7.5 x 2.67 = 20 miles + 5 miles = 25 miles out

g. If fuel consumption is 15.3 GPH and ground speed is 167 knots, how much fuel is required for an aircraft to travel 620 NM?

i.   620 nautical miles ÷ 167 knots = 221 minutes or
     3 hours and 41 minutes

ii. 15.3 GPH x 3 hrs. 41 min. = 57 gallons of fuel used

h. If the ground speed is 215 knots, how far will the aircraft travel in 3 minutes?

    i. 215 knots ÷ 60 = 3.58 nautical miles per minute

    ii. 3.58 NMPM x 3 minutes = 10.75 nautical miles

i. How accurate should you consider the predictions of performance charts to be?

Flight tests from which performance data was obtained were flown with a new, clean airplane, correctly rigged and loaded, and with an engine capable of delivering its full rated power. You can expect to do as well only if your airplane, too, is kept in peak condition.

D. Weight and Balance

1. What performance characteristics will be adversely affected when an aircraft has been overloaded? (AC 61-23C)

a. Increased takeoff speed

b. Increased takeoff runway length

c. Reduced rate-of-climb

d. Maximum altitude capability reduced

e. Operational range reduced

f. Less maneuverability

g. Less controllability

h. Higher stall speed

i. Higher approach speed

j. Longer landing distance

2. If, due to the addition or removal of fixed equipment in the aircraft, the weight and balance of an aircraft has changed, what responsibility does the owner or operator have?

The owner or operator of the aircraft should insure that maintenance personnel make appropriate entries in the aircraft records when repairs or modifications have been accomplished. Weight changes must be accounted for and proper notations made in weight and balance records. The appropriate form for these changes is "Major Repairs and Alterations" (FAA Form 337).

3. Define the term "center of gravity." (AC 61-23C)

The center of gravity (CG) is the point about which an aircraft would balance if it were possible to support the aircraft at that point. It is the mass center of the aircraft, or the theoretical point at which the entire weight of the aircraft is assumed to be concentrated. The CG must be within specific limits for safe flight.

4. What effect does a forward center of gravity have on an aircraft's flight characteristics? (AC 61-23C)

Higher stall speed—Stalling angle of attack reached at a higher speed due to increased wing loading.

Slower cruise speed—Increased drag, greater angle of attack required to maintain altitude.

More stable—When angle of attack is increased, the airplane tends to reduce angle of attack; longitudinal stability.

Greater back elevator pressure required—Longer takeoff roll, higher approach speeds and problems with the landing flare.

5. What effect does an aft center of gravity have on an aircraft's flight characteristics? (AC 61-23C)

Lower stall speed—Less wing loading.

Higher cruise speed—Reduced drag, smaller angle of attack required to maintain altitude.

Less stable—Stall and spin recovery more difficult; when angle of attack is increased it tends to result in additional increased angle of attack.

6. Define the following:
Arm; Basic Operating Weight; Center of Gravity; Center of Gravity Limits; Center of Gravity Range; Datum; Empty Weight; Fuel Load; LEMAC; Maximum Allowable Zero Fuel Weight; Maximum Landing Weight, Maximum Takeoff Weight; Mean Aerodynamic Chord; Moment; Moment Index; Ramp or Taxi Weight; Station; Useful Load. (AC 61-23C)

Arm—The horizontal distance in inches from the reference datum line to the center of gravity of an item.

Basic Operating Weight—The weight of the aircraft, including the crew, ready for flight but without payload and fuel. This term is only applicable to transport aircraft.

Center of Gravity—The point about which an aircraft would balance if it were possible to suspend it at that point, expressed in inches from datum.

Center of Gravity Limits—The specified forward and aft or lateral points beyond which the CG must not be located during takeoff, flight or landing.

Center of Gravity Range—The distance between the forward and aft CG limits indicated on pertinent aircraft specifications.

Datum—An imaginary vertical plane or line from which all measurements of arm are taken. It is established by the manufacturer.

Empty Weight—The airframe, engines, and all items of operating equipment that have fixed locations and are permanently installed in the aircraft. It includes optional and special equipment, fixed ballast, hydraulic fluid, unusable fuel, and undrainable oil.

Fuel Load—The expendable part of the load of the aircraft. It includes only usable fuel, not fuel required to fill the lines or that which remains trapped in the tank sumps.

LEMAC—The leading edge of the mean aerodynamic chord.

Continued

Maximum Allowable Zero Fuel Weight—The maximum weight authorized for the aircraft not including fuel load. Zero fuel weight for each particular flight is the operating weight plus the payload.

Maximum Landing Weight—The maximum weight at which the aircraft may normally be landed. The maximum landing weight may be limited to a lesser weight when runway length or atmospheric conditions are adverse.

Maximum Takeoff Weight—The maximum allowable weight at the start of the takeoff run. Some aircraft are approved for loading to a greater weight (ramp or taxi) only to allow for fuel burnoff during ground operation. The takeoff weight for a particular flight may be limited to a lesser weight when runway length, atmospheric conditions, or other variables are adverse.

Mean Aerodynamic Chord (MAC)—The average distance from the leading edge to the trailing edge of the wing. The MAC is specified for the aircraft by determining the average chord of an imaginary wing which has the same aerodynamic characteristics as the actual wing.

Moment—The product of the weight of an item multiplied by its arm. Moments are expressed in pound-inches.

Moment Index—A moment divided by a constant such as 100, 1,000, or 10,000. The purpose of using a moment index is to simplify weight and balance computations of large aircraft where heavy items and long arms result in large, unmanageable numbers.

Ramp or Taxi Weight—The maximum takeoff gross weight plus fuel to be burned during taxi and runup.

Station—A location in the aircraft which is identified by a number designating its distance in inches from the datum. The datum is, therefore, identified as station zero. The station and arm are usually identical. An item located at station +50 would have an arm of 50 inches.

Useful Load—The weight of the pilot, copilot, passengers, baggage, usable fuel and drainable oil. It is the empty weight subtracted from the maximum allowable takeoff weight. The term applies to general aviation aircraft only.

7. **What basic equation is used in all weight and balance problems to find the center of gravity location of an airplane and/or its components?** (AC 61-23C)

Weight x Arm = Moment

By rearrangement of this equation to the forms,

   Weight = Moment ÷ arm.

   Arm = Moment ÷ weight.

   CG = Moment ÷ weight.

With any two known values, the third value can be found.

8. **What basic equation is used to determine center of gravity?** (AC 61-23C)

Center of gravity is determined by dividing total moments by total weight.

9. **Explain the term "percent of mean aerodynamic chord (MAC)."** (AC 61-23C)

Expression of the CG relative to the MAC is a common practice in larger aircraft. The CG position is expressed as a percent MAC (percent of mean aerodynamic chord), and the CG limits are expressed in the same manner. Normally, an aircraft will have acceptable flight characteristics if the CG is located somewhere near the 25% average chord point. This means the CG is located one-fourth of the total distance back from the leading edge of the average wing section.

10. **If the weight of an aircraft is within takeoff limits but the CG limit has been exceeded, what actions can the pilot take to correct the situation?** (AC 61-23C)

The most satisfactory solution to this type of problem is to shift baggage, passengers, or both in an effort to make the aircraft CG fall within limits.

11. When a shift in weight is required, what standardized and simple calculations can be made to determine the new CG? (AC 61-23C)

A typical problem may involve calculation of a new CG for an aircraft which has shifted cargo due to the CG being out of limits.

Given:

Aircraft total weight 6,680

CG .... Station 80.0

CG limits .... Station 70-78

Find: What is the location of the CG if 200 pounds is shifted from the aft compartment at station 150 to the forward at station 30?

Solution:

a. $\dfrac{\text{Weight shifted x Distance moved}}{\text{Aircraft gross weight}}$ = CG change

b. $\dfrac{200 \times 120}{6,680}$ = 3.6 inches forward

c. old CG   80.0 inches
   minus change   − 3.6
   new CG   76.4 inches

This same formula may be used to calculate how much weight must be shifted when you know how far you want to move the CG to come within limits.

12. If the weight of an aircraft changes due to the addition or removal of cargo or passengers before flight, what formula may be used to calculate new CG? (AC 61-23C)

A typical problem may involve the calculation of a new CG for an aircraft which, when loaded and ready for flight, receives some additional cargo or passengers just before departure time.

Given:

Aircraft total weight .... 6,860 lbs.

CG .... Station 80.0

Find: What is the location of the CG if 140 lbs. of baggage is added to station 150?

Solution:

a. Use the added weight formula:

$$\frac{\text{Added weight (or removed)}}{\text{New total weight}} = \frac{\text{CG change}}{\text{Distance between weight \& old CG}}$$

$$\frac{140}{6,860 + 140} = \frac{\text{CG change}}{150 - 80}$$

$$\frac{140}{7,000} = \frac{\text{CG change}}{70}$$

CG change = 1.4 inches aft

b. Add the CG change to the old CG:
New CG = 80.0 in. + 1.4 in. = 81.4 in.

By using "old total weight and new" CG, this same formula may be used to find out how much weight to add or remove, when it is known how far you want to move the CG to come within limits.

13. What simple and fundamental weight check can be made by all pilots before flight? (AC 61-23C)

A useful load check can be made to determine if the useful load limit has been exceeded. This check may be a mental calculation if the pilot is familiar with the aircraft's limits and knows that unusually heavy loads are not aboard. The pilot needs to know the useful load limit of the particular aircraft. This information may be found in the latest weight and balance report, in a logbook, or on a Major Repair and Alteration Form located in the aircraft. If the useful load limit is not stated directly, simply subtract the empty weight from the maximum takeoff weight.

14. What factors would contribute to a change in center of gravity location during flight? (AC 61-23C)

The operator's flight manual should show procedures which fully account for the extreme variations in CG travel during flight caused by all of the following variables:

a. The movement of passengers;

b. Possible change in CG position due to landing gear retraction; and

c. The effect of the CG travel during flight due to fuel used.

15. If actual weights for weight and balance computations are unknown, what weights may be assumed for weight and balance computations? (AC 61-23C)

Some standard weights used in general aviation are:

Crew and Passengers .... 170 lbs./person

Gasoline .... 6 lbs./U.S. gal.

Oil .... 7.5 lbs./U.S. gal.

Water .... 8.35 lbs./U.S. gal.

Note: These weights are not to be used in lieu of available actual weights!

16. How is the CG affected during flight as fuel is used? (AC 61-23C)

As fuel is burned during flight, the weight of the fuel tanks will change and as a result the CG will change. Most aircraft, however, are designed with the fuel tanks positioned close to the CG; therefore, the consumption of fuel does not affect the CG to any great extent. Generally, larger aircraft require more careful weight and balance computations due to the increased amounts of fuel being carried and the possibility of the CG shifting out of limits as that fuel is burned.

Weight and Balance Review

1. What is the center of gravity range for this aircraft at maximum takeoff weight?

2. What are the empty, basic, and maximum takeoff weights of your aircraft?

3. What is the maximum allowable weight for the baggage compartment?

4. How many people will this aircraft carry safely with a full fuel load?

5. What is your maximum allowable useful load?

6. Solve a weight and balance problem for the flight you plan to make with one FAA examiner on board. (Be sure to ask their actual weight!)

   a. Does your load fall within the weight and balance envelope?

   b. What is the final gross weight?

   c. How much fuel can be carried?

   d. How much baggage can be carried with full fuel?

7. Calculate the new CG for the above aircraft if an additional 80 pounds of baggage were loaded.

Additional Study Questions

1. What are three ways you can control lift during flight? (AC 61-23C)

2. At what altitude will V_X and V_Y be the same? (FAA-H-8083-3)

3. Define the terms "service ceiling" and "absolute ceiling." (FAA-H-8083-3)

4. Why does an airplane turn? (AC 61-23C)

5. What makes an airplane climb? (AC 61-23C)

6. Why will an airplane (T-tail excluded) pitch nosedown when power is reduced and no adjustment is made with the flight controls? (FAA-H-8083-3)

7. What causes the left turning tendency in a tricycle gear aircraft on the takeoff roll? (FAA-H-8083-3)

8. During enroute phase of flight, the CG moves aft due to fuel burn or a passenger moving. What effect will this have on your airspeed and stall speed? (AC 61-23C)

9. Discuss some of the other factors that will affect performance of an aircraft that are not included in the aircraft performance charts. (AC 61-23C)

10. What is the principle factor that determines propeller efficiency? (AC 61-23C)

Cross-Country Flight Planning and Procedures

6

A. Flight Plan

Be prepared to exhibit knowledge of the elements related to cross-country flight planning by presenting and explaining a pre-planned VFR cross-country flight, as previously assigned by the examiner. On the day of the test, the final flight plan shall include real-time weather to the first fuel stop. Computations shall be based on maximum allowable passenger, baggage and/or cargo loads.

The examiner will ask you to discuss the following:

1. Current and forecast weather at departure, en route, destination and alternate airports including:

a. Surface observations
b. Terminal forecasts
c. Area forecasts
d. Winds aloft forecasts
e. TWEB route forecasts
f. SIGMETs and AIRMETs
g. Pilot Reports

2. Route selection including:

a. Selection of checkpoints
b. Selection of best altitude
c. Selection of alternate airport

3. Appropriate sectional charts:

a. Knowledge of chart symbols
b. Airspace
c. Communication frequencies

4. Current information on facilities and procedures:

a. NOTAMs
b. Special Notices
c. Services available at destination
d. Airport conditions including lighting, obstructions, and other notations in AFD

5. Flight log:

a. Measurement of course (true and magnetic)
b. Distances between checkpoints and total
c. How true airspeed was obtained
d. Estimated ground speed
e. Total time en route
f. Amount of fuel required

6. Weight and balance:

a. Calculations for planned trip
b. Calculations for weight added or removed immediately before departure

B. Navigation

1. What is an RMI? (Pilot/Controller Glossary)

RMI is an abbreviation for radio magnetic indicator. It is an aircraft navigational instrument coupled with a gyro compass or similar compass that indicates the direction of a selected NAVAID (NDB or VOR) and indicates bearing with respect to the heading of the aircraft.

2. What is an HSI? (AC 61-27C)

HSI is an abbreviation for *horizontal situation indicator.* It is a combination of two instruments, the directional gyro and the VOR/ILS indicator. Combining the DG and NAV indicator into one instrument reduces pilot workload by providing heading, course reference, course deviation and glide slope information all into one visual aid.

3. What is RNAV? (Pilot/Controller Glossary)

RNAV is an abbreviation for *area navigation.* It is a form of navigation that permits aircraft properly equipped to operate on any desired course within the coverage of station-referenced navigation signals. This form of navigation allows a pilot to select a more direct course to a destination by not requiring overflight of ground-based navigational aids. Navigation is to selected "way-points" instead of VORs. A "waypoint" is created by simply moving the VOR to a point along the route of flight desired.

4. What is DME? (AIM 1-1-7)

Equipment (airborne and ground) used to measure, in nautical miles, the slant range distance of an aircraft from the DME navigational aid. Aircraft equipped with DME are provided with distance and groundspeed information when receiving a VORTAC or TACAN facility. DME operates on frequencies in the UHF spectrum between 962 MHz and 1213 MHz.

5. What is the effective range distance for DME? (AIM 1-1-7)

Operating on the line-of-sight principle, DME furnishes distance information with a very high degree of accuracy. Reliable signals may be received at distances up to 199 NM at line-of-sight altitude with an accuracy of better than ½ mile or 3 percent of the distance, whichever is greater. Distance information received from DME equipment is SLANT RANGE distance and not actual horizontal distance.

6. What is GPS? (AIM 1-1-22)

Global Positioning System (GPS)—A satellite-based navigation system which provides highly accurate position and velocity information, and precise time, on a continuous global basis to an unlimited number of properly equipped aircraft. The system is unaffected by weather, and provides a worldwide common grid reference system. The GPS concept is predicated upon accurate and continuous knowledge of the spatial position of each satellite in the system with respect to time and distance from a transmitting satellite to the aircraft. The GPS receiver automatically selects appropriate signals from the satellites in view and translates these into a three-dimensional position, velocity, and time. System accuracy for civil users is 100 meters horizontally.

7. Within which frequency band does the VOR equipment operate? (AIM 1-1-3)

VHF band—108.00 through 117.95 MHz

8. What are the different methods for checking the accuracy of VOR equipment? (14 CFR 91.171)

a. VOT check; ±4°
b. Ground checkpoint; ±4°
c. Airborne checkpoint; ±6°
d. Dual VOR check; 4° between each other
e. Select a radial over a known ground point; ±6°

A repair station can use a radiated test signal, but only the technician performing the test can make an entry in the logbook.

9. What records must be kept concerning VOR checks? (14 CFR 91.171)

Each person making a VOR check shall enter the date, place, and bearing error and sign the aircraft log or other reliable record.

10. Where can a pilot find the location of the nearest VOT testing stations? (AIM 1-1-4)

Locations of airborne check points, ground check points and VOTs are published in the A/FD and are depicted on the A/G voice communications panels on the NOS IFR area chart and IFR enroute low altitude chart.

11. How may the course sensitivity be checked on a VOR receiver? (AC 61-27C)

In addition to receiver tolerance checks required by regulations, course sensitivity may be checked by recording the number of degrees of change in the course selected as you rotate the OBS to move the CDI from center to the last dot on either side. This should be between 10° and 12°.

12. How can a pilot determine if a VOR or VORTAC has been taken out of service for maintenance? (AIM 1-1-12)

During periods of routine or emergency maintenance, coded identification (or code and voice, where applicable) is removed from certain FAA NAVAIDs. Removal of identification serves as a warning to pilots that the facility is officially off the air for tune-up or repair and may be unreliable even though intermittent or constant signals are received.

13. How do you find an ADF "relative bearing"? (AC 61-27C)

A relative bearing is the angular relationship between the aircraft heading and the station measured clockwise from the nose. The bearing is read directly on the ADF dial measured clockwise from zero.

14. How do you find an ADF "magnetic bearing"? (AC 61-27C)

A magnetic bearing is the direction of an imaginary line from the aircraft to the station or the station to the aircraft referenced to magnetic north. To determine, use this formula:

MH + RB = MB

(Magnetic Heading + Relative Bearing = Magnetic Bearing)

If the sum is more than 360, subtract 360 to get the magnetic bearing to the station. The reciprocal of this number is the magnetic bearing from the station.

15. What is ADF homing? (AC 61-27C)

ADF homing is flying the aircraft on any heading required to keep the ADF needle on zero until the station is reached.

16. What is ADF tracking? (AC 61-27C)

ADF tracking is a procedure used to fly a straight geographic flight path inbound to or outbound from an NDB. A heading is established that will maintain the desired track.

17. **If a diversion to an alternate airport becomes necessary due to an emergency, what procedure should be used?** (AC 61-23C)

    a. Consider relative distance to all suitable alternates;

    b. Select the one most appropriate for the emergency at hand;

    c. Determine magnetic course to alternate and divert immediately;

    d. Wind correction, actual distance and estimated time/fuel can then be computed while enroute to alternate.

18. **How can the course to an alternate be computed quickly?** (AC 61-23C)

    Courses to alternates can be quickly measured by using a straight edge and the compass roses shown at VOR stations on the chart. VOR radials and airway courses (already oriented to magnetic direction) printed on the chart can be used to approximate magnetic bearings during VFR flights. Use the radial of a nearby VOR or airway that most closely parallels the course to the station. Distances can be determined by placing a finger at the appropriate place on a straight edge of a piece of paper and then measuring the approximate distance on the mileage scale at the bottom of the chart.

19. **How do you determine time and distance from a VOR/NDB station?** (AC 61-27C)

    a. Determine the radial on which you are located.

    b. Turn 80° right or left of the inbound course rotating the OBS to the nearest 10° increment opposite the direction of turn.

    c. Maintain heading. When the CDI centers, note the time.

    d. Maintaining the same heading, rotate the OBS 10° in the same direction as above.

    e. Note the elapsed time when the CDI again centers.

    f. Time/distance from the station is determined from the following formulas:

Time to station:

$$\frac{\text{Time in seconds between bearing change}}{\text{Degrees of bearing change}}$$

Distance to station:

$$\frac{\text{TAS x minutes flown}}{\text{Degrees of bearing change}}$$

20. **After 150 miles are flown from the departure point the aircraft's position is located 8 miles off course. If 160 miles remain to be flown, what approximate total correction should be made to converge on the destination?**

$$\frac{\text{Miles Off Course}}{\text{Total Miles Flown}} \text{ x } 60 = \text{Degrees to Parallel}$$

$$\frac{\text{Miles Off Course}}{\text{Total Miles Remaining}} \text{ x } 60 = \text{Degrees to Intercept}$$

Add the degrees to parallel and the degrees to intercept for the total correction to re-intercept course.

a. (8 miles off course ÷ 150) x 60 = 3.2°

b. (8 miles off course ÷ 160) x 60 = 3.0°

c. 3.2° to parallel + 3.0° to intercept = 6.2° to converge.

C. Airport and Traffic Pattern Operations

1. **What recommended entry and departure procedures should be utilized at airports without an operating control tower?** (AIM 4-3-3)

When entering a traffic pattern, enter the pattern in level flight, abeam the midpoint of the runway at pattern altitude. Maintain pattern altitude until abeam the approach end of the landing runway on the downwind leg. Complete the turn to final at least ¼ mile from the runway. When departing a traffic pattern, continue straight out, or exit with a 45-degree turn (to the left when in a left-hand traffic pattern; to the right when in a right-hand traffic pattern) beyond the departure end of the runway, after reaching pattern altitude.

2. **What are the recommended traffic advisory practices at airports *without* an operating control tower?** (AIM 4-1-9)

Pilots of inbound traffic should monitor and communicate as appropriate on the designated CTAF from 10 miles to landing. Pilots of departing aircraft should monitor/communicate on the appropriate frequency from start-up, during taxi, and until 10 miles from the airport unless federal regulations or local procedures require otherwise.

3. **A large or turbine-powered aircraft is required to enter Class D airspace at what altitude?** (14 CFR 91.129)

A large or turbine-powered airplane shall, unless otherwise required by the applicable distance-from-clouds criteria, enter the traffic pattern at an altitude of at least 1,500 feet above the elevation of the airport and maintain at least 1,500 feet until further descent is required for a safe landing.

4. **If operating into an airport without an operating control tower which is located within the Class D airspace of an airport with an operating control tower, is it always necessary to communicate with the tower?** (14 CFR 91.129)

Yes, operations to or from an airport in Class D airspace (airport traffic area) require communication with the tower even when operating to/from a satellite airport.

5. **When conducting flight operations into an airport with an operating control tower, when should initial contact be established?** (AIM 4-3-2)

When operating at an airport where traffic control is being exercised by a control tower, pilots are required to maintain two-way radio contact with the tower while operating within Class B, Class C, and Class D surface areas unless the tower authorizes otherwise. Initial call-up should be made about 15 miles from the airport.

6. **When departing a Class D surface area, what communication procedures are recommended?** (AIM 4-3-2)

   Unless there is good reason to leave the tower frequency before exiting the Class B, Class C and Class D surface areas, it is good operating practice to remain on the tower frequency for the purpose of receiving traffic information. In the interest of reducing tower frequency congestion, pilots are reminded that it is not necessary to request permission to leave the tower frequency once outside of Class B, Class C, and Class D surface areas.

7. **You discover that both the transmitter and receiver in your aircraft have become inoperative. What procedures should be used when attempting to enter a traffic pattern and land at a tower controlled airport?** (AIM 4-2-13)

   a. Remain outside or above Class D surface area.

   b. Determine direction and flow of traffic.

   c. Join the traffic pattern and wait for light gun signals.

   d. Daytime, acknowledge by rocking wings. Nighttime, acknowledge by flashing landing light or navigation lights.

8. **When a control tower, located at an airport within Class D airspace, ceases operation for the day, what happens to the lower limit of the controlled airspace?** (AIM 3-2-5)

   During the hours the tower is not in operation, Class E surface area rules or a combination of Class E rules down to 700 feet AGL and Class G rules to the surface will become applicable. Check the A/FD for specifics.

9. **If the rotating beacon is on at an airport during daylight hours, what significance does this have?** (AIM 2-1-8)

   In Class B, Class C, Class D, and Class E surface areas, operation of the airport beacon during the hours of daylight often indicates that the ground visibility is less than 3 miles and/or the ceiling is less than 1,000 feet. ATC clearance in accordance with Part 91 is required for landing, takeoff and flight in the traffic pattern. Pilots

Continued

should not rely solely on the operation of the airport beacon to indicate if weather conditions are IFR or VFR. There is no regulatory requirement for daylight operation, and it is the pilot's responsibility to comply with proper preflight planning as required by 14 CFR Part 91.

10. What are the various types of runway markings (precision instrument runway) and what do they consist of? (AIM 2-3-3)

Markings for runways are white and consist of the following types:

a. *Runway designators*—runway number is the whole number nearest one tenth the magnetic azimuth of the centerline of the runway, measured clockwise from magnetic north.

b. *Runway centerline marking*—identifies center of runway providing alignment and guidance during takeoff and landings; consists of a line of uniformly spaced stripes and gaps.

c. *Runway threshold markings*—identifies the beginning of the runway that is available for landing. In some instances the landing threshold may be relocated or displaced. The markings come in two configurations; either eight longitudinal stripes of uniform dimensions disposed symmetrically about the runway centerline or the number of stripes is related to the runway width.

d. *Runway aiming point marking*—serves as a visual aiming point for a landing aircraft; two rectangular markings consist of a broad white stripe located on each side of the runway centerline and approximately 1,000 feet from the landing threshold.

e. *Runway touchdown zone markers*—identify touch down zone for landing operations and are coded to provide distance information in 500 feet increments; consists of groups of one, two and three rectangular bars symmetrically arranged in pairs about the runway centerline.

11. What are the various types of taxiway markings and what do they consist of? (AIM 2-3-4)

Markings for taxiways are yellow and consist of the following types:

a. *Taxiway centerline*—single continuous yellow line; provides wingtip clearance when over center.

b. *Taxiway edge*—used to define the edge of taxiway; two types: continuous and dashed.

c. *Taxiway shoulder*—usually defined by taxiway edge markings; denotes pavement unusable for aircraft.

d. *Surface painted taxiway direction*—yellow background with black inscription; supplements direction signs or when not possible to provide taxiway sign.

e. *Surface painted location signs*—black background with yellow inscription; supplements location signs.

f. *Geographic position markings*—located at points along low visibility taxi routes; used to identify aircraft during low visibility operations.

12. What are the six types of signs installed on airports? (AIM 2-3-7)

a. *Mandatory instruction signs*—red background/white inscription; denotes hazardous areas.

b. *Location signs*—black background/yellow inscription; used to identify either a taxiway or runway on which an aircraft is located.

c. *Direction signs*—yellow background/black inscription; identifies designation (s) of intersecting taxiway (s) leading out of intersection that pilot would expect to turn onto or hold short of.

d. *Destination signs*—yellow background/ black inscription; signs have arrow showing direction of taxi route to that destination.

e. *Information signs*—yellow background/ black inscription; provide pilot information on such things as areas that cannot be seen by control tower, radio frequencies, noise abatement procedures, etc.

f. *Runway distance remaining signs*—black background with white numeral inscription; indicates distance (in thousands of feet) of landing runway remaining.

13. What is LAHSO? (AIM 4-3-11)

An acronym for "Land and Hold Short Operations." These include landing and holding short of an intersecting runway, an intersecting taxiway, or some other designated point on a runway. LAHSO is an ATC procedure that requires pilot participation to balance the needs for increased airport capacity and system efficiency.

14. When should you decline a LAHSO clearance? (AIM 4-3-11)

Student pilots or pilots not familiar with LAHSO should not participate in the program. Pilots are expected to decline a LAHSO clearance if they determine it will compromise safety or if weather is below basic VFR weather conditions (a minimum ceiling of 1,000 feet and 3 statute miles visibility).

15. Describe a "displaced threshold." (AIM 2-3-3)

A displaced threshold is a threshold located at a point on the runway other than the designated beginning of the runway. Displacement of the threshold reduces the length of the runway available for landings. The portion of the runway behind a displaced threshold is available for takeoffs in either direction and landings from the opposite direction. A ten-foot-wide white threshold bar is located across the width of the runway at the displaced threshold. White arrows are located along the centerline in the area between the beginning of the runway and the displaced threshold. White arrowheads are located across the width of the runway just prior to the threshold bar.

16. Describe a tri-color light VASI system. (AIM 2-1-2)

A tri-color Visual Approach Slope Indicator (VASI) normally consists of a single light unit projecting a three color visual approach path into the final approach area of the runway.

Red ....Below glidepath
Amber ....Above glidepath
Green ....On glidepath

17. What is "PAPI"? (AIM 2-1-2)

The Precision Approach Path Indicator (PAPI) uses light units similar to the VASI but are installed in a single row of either two or four light units. These systems have an effective visual range of about 5 miles during the day and up to 20 miles at night. The row of light units is normally installed on the left side of the runway.

18. What is "PLASI"? (AIM 2-1-2)

A Pulse Light Approach Slope Indicator (PLASI) is a pulsating visual approach slope indicator that normally consists of a single light unit projecting a two-color visual approach path into the final approach area. Effective range for this system is 4 miles in the daytime and up to 10 miles at night.

Pulsating white light ....Above glidepath

Steady white light ....On glidepath

Steady red light ....Slightly below glidepath

Pulsating red light ....Well below glidepath

D. 14 CFR Part 91

1. Can a commercial pilot allow a passenger to carry alcohol on board an aircraft for the purpose of consumption? (14 CFR 91.17)

No, the regulations do not specifically address this issue but do indicate that a person who is intoxicated (or becomes intoxicated) not be allowed on board an aircraft. Except in an emergency, no pilot of a civil aircraft may allow a person who appears to be intoxicated or who demonstrates by manner or physical indications that the individual is under the influence of drugs (except a medical patient under proper care) to be carried in that aircraft.

2. No person may act as a crewmember of a civil aircraft with a blood alcohol level of what value? (14 CFR 91.17)

No person may act or attempt to act as a crewmember of a civil aircraft while having .04% by weight or more alcohol in the blood.

3. When are the operation of portable electronic devices not allowed on board an aircraft? (14 CFR 91.21)

No person may operate, nor may any operator or pilot-in-command of an aircraft allow the operation of any portable electronic device on any of the following U.S.-registered aircraft:

a. Aircraft operated by a holder of an air carrier operator certificate or an operating certificate, or

b. Any other aircraft while it is operated under IFR.

4. Are there any exceptions allowed concerning portable electronic equipment on board aircraft? (14 CFR 91.21)

a. Portable voice recorders

b. Hearing aids

c. Heart pacemakers

d. Electric shavers

e. Any other portable electronic device that the operator of the aircraft has determined will not cause interference with the navigation or communication system of the aircraft on which it is to be used.

5. Preflight action as required by regulation for all flights away from the vicinity of the departure airport shall include a review of what specific information? (14 CFR 91.103)

For a flight under IFR or a flight not in the vicinity of an airport:

a. Weather reports and forecasts

b. Fuel requirements

c. Alternatives available if the planned flight cannot be completed

d. Any known traffic delays of which the pilot-in-command has been advised by ATC

e. Runway lengths of intended use

f. Takeoff and landing distance data

6. When are flight crewmembers required to wear their seatbelts? (14 CFR 91.105)

During takeoff and landing, and while en route, each required flight crewmember shall keep the safety belt fastened while at the crewmember station (also, during takeoff and landing only, the shoulder harness, if installed).

7. Is the use of safety belts and shoulder harnesses required when operating an aircraft on the ground? (14 CFR 91.107)

Yes; each person on board a U.S.-registered civil aircraft must occupy an approved seat or berth with a safety belt, and if installed, shoulder harness, properly secured about him or her during movement on the surface, takeoff and landing.

8. If a formation flight has been arranged in advance, can passengers be carried for hire? (14 CFR 91.111)

No; no person may operate an aircraft, carrying passengers for hire, in formation flight.

9. What is the maximum speed allowed when operating inside Class B airspace, under 10,000 feet and within a Class D surface area? (14 CFR 91.117)

Unless otherwise authorized or required by ATC, no person may operate an aircraft at or below 2,500 feet above the surface within 4 nautical miles of the primary airport of a Class C or Class D airspace area at an indicated airspeed of more than 200 knots. This restriction does not apply to operations conducted within a Class B airspace area. Such operations shall comply with the "below 10,000 feet MSL" restriction:

"No person shall operate an aircraft below 10,000 feet MSL, at an indicated airspeed of more than 250 knots."

10. What regulations pertain to altimeter setting procedures? (14 CFR 91.121)

Below 18,000 feet MSL:

a. The current reported altimeter setting of a station along the route and within 100 nautical miles of the aircraft.

b. If there is no station within the area described above, the current reported altimeter of an appropriate available station.

c. In the case of an aircraft not equipped with a radio, the elevation of the departure airport or an appropriate altimeter setting available before departure.

Note: If barometric pressure exceeds 31.00" Hg, set 31.00" (*see* AIM).

At or above 18,000 feet MSL set to 29.92" Hg.

11. Briefly describe the six classes of U.S. airspace. (AIM 3-2-2 through 3-2-6, and 3-3-1)

Class A airspace—Generally, airspace from 18,000 feet MSL up to and including FL600, including airspace overlying the waters within 12 nautical miles of the coast of the 48 contiguous states and Alaska; and designated international airspace beyond 12 nautical miles of the coast of the 48 contiguous states and Alaska within areas of domestic radio navigational signal or ATC radar coverage, and within which domestic procedures are applied.

Class B airspace—Generally, airspace from the surface to 10,000 feet MSL surrounding the nation's busiest airports in terms of IFR operations or passenger enplanements. The configuration of each Class B airspace area is individually tailored and consists of a surface area and two or more layers, (some resemble upside-down wedding cakes), and is designated to contain all published instrument procedures once an aircraft enters the airspace. An ATC clearance is required for all aircraft to operate in the area, and all aircraft cleared as such receive separation services within the airspace. The cloud clearance requirement for VFR operations is "clear of clouds."

Class C airspace—Generally, airspace from the surface to 4,000 feet above the airport elevation (charted in MSL) surrounding airports that have an operational control tower, are serviced by a radar approach control, and that have a certain number of IFR operations or passenger enplanements. Although the configuration of each Class C airspace area is individually tailored, the airspace usually consists of a 5 NM radius core surface area that extends from the surface up to 4,000 feet above the airport elevation, and a 10 NM radius shelf area that extends from 1,200 feet to 4,000 feet above the airport elevation.

Class D airspace—Generally, airspace from the surface to 2,500 feet above the airport elevation (charted in MSL) surrounding airports that have an operational control tower. The configuration of each Class D airspace area is individually tailored and when instrument procedures are published, the airspace will normally be designed to contain those procedures.

Class E (controlled) airspace—Generally, if the airspace is not Class A, Class B, Class C, or Class D, and it is controlled airspace, it is Class E airspace. Class E airspace extends upward from either the surface or a designated altitude to the overlying or adjacent controlled airspace. Examples of Class E airspace include: Surface areas designated for an airport, extensions to a surface area, airspace used for transition, enroute domestic areas, Federal airways, offshore airspace areas.

Class G (uncontrolled) airspace—Class G airspace is that portion of the airspace that has not been designated as Class A, Class B, Class C, Class D, and Class E airspace.

12. What are the regulatory fuel requirements for both VFR and IFR flight (day and night)? (14 CFR 91.151, 91.167)

a. *VFR conditions:*

No person may begin a flight in an airplane under VFR conditions unless (considering wind and forecast weather conditions) there is enough fuel to fly to the first point of intended landing and, assuming normal cruising speed:

    i. During the day, to fly after that for at least 30 minutes; or
    ii. At night, to fly after that for at least 45 minutes.

Continued

b. *IFR conditions:*

No person may operate a civil aircraft in IFR conditions unless it carries enough fuel (considering weather reports and forecasts) to:

  i. Complete the flight to the first airport of intended landing;
  ii. Fly from that airport to the alternate airport; and
  iii. Fly after that for 45 minutes at normal cruising speed.

If an alternate is not required, complete the flight to the destination airport with a 45-minute reserve remaining.

13. When conducting IFR flight operations, what minimum altitudes are required over surrounding terrain? (14 CFR 91.177)

If no applicable minimum altitudes apply:

a. Operations over an area designated as a mountainous area, an altitude of 2,000 feet above the highest obstacle within a horizontal distance of 4 nautical miles from the course to be flown; or

b. In any other case, an altitude of 1,000 feet above the highest obstacle within a horizontal distance of 4 nautical miles from the course to be flown.

14. What are several examples of situations in which an ELT is not required equipment on board the aircraft? (14 CFR 91.207)

Examples of operations where an ELT is not required are:

a. Ferrying aircraft for installation of an ELT

b. Ferrying aircraft for repair of an ELT

c. Aircraft engaged in training flights within a 50-nautical mile radius of an airport.

15. Where is altitude encoding transponder equipment required? (AIM 4-1-19)

In general, the regulations require aircraft to be equipped with Mode C transponders when operating:

a. At or above 10,000 feet MSL over the 48 contiguous states or the District of Columbia, excluding airspace below 2,500 feet AGL;

b. Within 30 miles of a Class B airspace primary airport, below 10,000 feet MSL;

c. Within and above all Class C airspace, up to 10,000 feet MSL;

d. Within 10 miles of certain designated airports, excluding airspace which is both outside the Class D surface area and below 1,200 feet AGL;

e. All aircraft flying into, within, or across the contiguous United States ADIZ.

16. Where are aerobatic flight maneuvers not permitted? (14 CFR 91.303)

No person may operate an aircraft in aerobatic flight—

a. Over any congested area of a city, town, or settlement;

b. Over an open air assembly of persons;

c. Within the lateral boundaries of the surface areas of Class B, Class C, Class D, or Class E airspace designated for an airport;

d. Within 4 nautical miles of the center line of any Federal airway;

e. Below an altitude of 1,500 feet above the surface; or

f. When flight visibility is less than 3 statute miles.

For the purposes of this section, aerobatic flight means an intentional maneuver involving an abrupt change in an aircraft's attitude, and abnormal attitude, or abnormal acceleration, not necessary for normal flight.

17. When must each occupant of an aircraft wear an approved parachute? (14 CFR 91.307)

a. Unless each occupant of the aircraft is wearing an approved parachute, no pilot of a civil aircraft carrying any person (other than a crewmember) may execute any intentional maneuver that exceeds:

   i. A bank of 60° relative to the horizon; or

   ii. A nose-up or nose-down attitude of 30° relative to the horizon.

Continued

b. This regulation does not apply to:

    i. Flight tests for pilot certification or rating; or

    ii. Spins and other flight maneuvers required by the regulations for any certificate or rating when given by a certified flight instructor or an Airline Transport Pilot.

18. What is required to operate an aircraft towing an advertising banner? (14 CFR 91.311)

No pilot of a civil aircraft may tow anything with that aircraft (other than under 91.309 "Towing gliders") except in accordance with the terms of a certificate of waiver issued by the Administrator.

19. What categories of aircraft cannot be used in the carriage of persons or property for hire? (14 CFR 91.313, 91.315, and 91.319)

a. Restricted category
b. Limited category
c. Experimental

E. AIM (Aeronautical Information Manual)

1. What is "primary radar" and "secondary radar"? (Pilot/Controller Glossary)

Primary radar—A radar system in which a minute portion of a radio pulse transmitted from a site is reflected by an object and then received back at that site for processing and display at an Air Traffic Control facility.

Secondary radar—A radar system in which the object to be detected is fitted with a transponder. Radar pulses transmitted from the searching transmitter/receiver (interrogator) site are received in the transponder and used to trigger a distinctive transmission from the transponder. The reply transmission, rather than the reflected signal, is then received back at the transmitter/receiver site for processing and display at an Air Traffic Control facility.

2. What is airport surveillance radar?
(Pilot/Controller Glossary)

Airport surveillance radar (ASR) is approach control radar used to detect and display an aircraft's position in the terminal area. ASR provides range and azimuth information but does not provide elevation data. Coverage of ASR can extend up to 60 miles.

3. Describe the various types of terminal radar services available for VFR aircraft. (AIM 4-1-17)

Basic radar service—Safety alerts, traffic advisories, limited radar vectoring (on a workload-permitting basis) and sequencing at locations where procedures have been established for this purpose and/or when covered by a letter of agreement.

TRSA service—Radar sequencing and separation service for participating VFR aircraft in a TRSA.

Class C service—This service provides, in addition to basic radar service, approved separation between IFR, and VFR aircraft, and sequencing of VFR arrivals to the primary airport.

Class B service—Provides, in addition to basic radar service, approved separation of aircraft based on IFR, VFR, and/or weight, and sequencing of VFR arrivals to the primary airport(s).

4. What frequencies are monitored by most FSS's other than 121.5? (AIM 4-2-14)

FSS's and supplemental Weather Service Locations are allocated frequencies for different functions: for example, 122.0 MHz is assigned as the Enroute Flight Advisory Service frequency at selected FSS's. In addition, certain FSS's provide Local Airport Advisory on 123.6 MHz. Frequencies are listed in the Airport/ Facility Directory. If you are in doubt as to what frequency to use, 122.2 MHz is assigned to the majority of FSS's as a common enroute simplex frequency.

5. If operations are not being conducted in airspace requiring a transponder, can an aircraft equipped with a transponder leave it off? (AIM 4-1-19)

In all cases, while in controlled airspace (Class A, B, C, D, or E airspace) each pilot operating an aircraft equipped with an operable ATC transponder maintained in accordance with 14 CFR §91.413 shall operate the transponder, including Mode C if installed, on the appropriate code or as assigned by ATC. In Class G airspace (uncontrolled airspace), the transponder should be operating while airborne, unless otherwise requested by ATC.

6. At what altitude would a pilot expect to encounter military aircraft when navigating through a military training route designated "VR1207"? (AIM 3-5-2)

Less than 1,500 AGL; Military training routes with no segment above 1,500 feet AGL shall be identified by four-digit characters; e.g., IR1206, VR1207. MTRs that include one or more segments above 1,500 feet AGL shall be identified by three-digit characters; e.g., IR206, VR207.

7. What is a "composite flight plan"? (AIM 5-1-6)

Flight plans which specify VFR operations for one portion of the flight and IFR for another portion will be accepted by the FSS at the point of departure. If VFR flight is conducted for the first portion and IFR for the last portion:

a. The pilot should report the departure time to the FSS with which he filed his VFR/IFR flight plan;

b. At the point of intended change, close the VFR portion;

c. Request ATC clearance from the FSS nearest the point at which the change from VFR to IFR is proposed; and

d. Remain in VFR weather conditions until operating in accordance with the IFR clearance.

8. What is an "abbreviated" IFR flight plan?
(PIlot/Controller Glossary)

An abbreviated IFR flight plan is an authorization by ATC requiring pilots to submit only that information needed for the purpose of ATC. It is frequently used by aircraft which are airborne and desire an instrument approach or by an aircraft on the ground which desires to climb to VFR-On-Top conditions.

9. How long will a flight plan remain on file after the proposed departure time has passed? (AIM 5-1-11)

To prevent computer saturation in the en route environment, parameters have been established to delete proposed departure flight plans which have not been activated. Most centers have this parameter set so as to delete these flight plans a minimum of 1 hour after the proposed departure time.

10. If you fail to report a change in arrival time or forget to close your flight plan, when will search and rescue procedures begin? (AIM 5-1-12)

If you fail to report or cancel your flight plan within ½ hour after your ETA, search and rescue procedures are started.

11. What constitutes a change in flight plan? (AIM 5-1-10)

In addition to altitude or flight level, destination and/or route changes, increasing or decreasing the speed of the aircraft constitutes a change in flight plan. Therefore, anytime the average true airspeed at cruising altitude between reporting points varies or is expected to vary from that given in the flight plan by ±5 percent or 10 knots, whichever is greater, ATC should be advised.

12. What is a DVFR flight plan? (AIM 5-1-5)

Defense VFR; VFR flights into a coastal or domestic ADIZ/DEWIZ are required to file VFR flight plans for security purposes. The flight plan must be filed before departure.

13. What is an ADIZ? (AIM 5-6-1)

All aircraft entering domestic U.S. airspace from points outside must provide for identification prior to entry. To facilitate early identification of all aircraft in the vicinity of U.S. and international airspace boundaries, Air Defense Identification Zones (ADIZs) have been established.

14. What requirements must be satisfied prior to operations into, within or across an ADIZ? (AIM 5-6-1)

Operational requirements for aircraft operations associated with an ADIZ are as follows:

Flight plan — An IFR or DVFR flight plan must be filed with the appropriate aeronautical facility.

Two-way radio — An operating two-way radio is required.

Transponder — Aircraft must be equipped with an operable radar beacon transponder having altitude reporting (Mode C) capabilities. The transponder must be turned on and set to the assigned ATC code.

Position reports — For IFR flights, normal position reporting. For DVFR flights, an estimated time of ADIZ penetration must be filed at least 15 minutes prior to entry.

Aircraft position tolerances — Over land, a tolerance of ±5 minutes from the estimated time over a reporting point and within 10 NM from the centerline of an intended track over an estimated reporting point. Over water, a tolerance of ±5 minutes from the estimated time over a reporting point or point of penetration and within 20 NM from centerline of an intended track over an estimated reporting point.

15. Define the following types of airspace.
(AIM 3-4-1 through 3-4-8)

Prohibited Area — For security or other reasons, aircraft flight is prohibited.

Restricted Area — Contains unusual, often invisible hazards to aircraft, flights must have permission from the controlling agency, if VFR. IFR flights will be cleared through or vectored around it.

Military Operations Area—Designed to separate military training from IFR traffic. Permission is not required, but VFR flights should exercise caution. IFR flights will be cleared through or vectored around it.

Warning Area—Same hazards as a restricted area, it is established beyond the 3-mile limit of International Airspace. Permission is not required, but a flight plan is advised.

Alert Area—Airspace containing a high volume of pilot training or unusual aerial activity. No permission is required, but VFR flights should exercise caution. IFR flights will be cleared through or vectored around it.

Controlled Firing Areas—CFAs contain activities which, if not conducted in a controlled environment, could be hazardous to non-participating aircraft. The distinguishing feature of the CFA, as compared to other special use airspace, is that its activities are suspended immediately when spotter aircraft, radar or ground lookout positions indicate an aircraft might be approaching the area. CFAs are not charted.

National Security Areas—Airspace of defined vertical and lateral dimensions established at locations where there is a requirement for increased security and safety of ground facilities. Pilots are requested to voluntarily avoid flying through the depicted NSA. When it is necessary to provide a greater level of security and safety, flight in NSAs may be temporarily prohibited by regulation under the provisions of 14 CFR §99.7.

16. What procedures should be utilized in avoiding wake turbulence when landing? (AIM 7-3-6)

a. Landing behind a larger aircraft, on the same runway: stay at or above the larger aircraft's final approach flight path. Note its touchdown point and land beyond it.

b. Landing behind a larger aircraft, on a parallel runway closer than 2,500 feet: consider possible drift to your runway. Stay at or above the larger aircraft's final approach flight path and note its touchdown point.

c. Landing behind a larger aircraft on a crossing runway: cross above the larger aircraft's flight path.

Continued

d. Landing behind a departing larger aircraft on the same runway: note the larger aircraft's rotation point, and land well before the rotation point.

e. Landing behind a departing larger aircraft on a crossing runway: note the larger aircraft's rotation point. If it is past the intersection, continue the approach, and land prior to the intersection. If the larger aircraft rotates prior to the intersection, avoid flight below the larger aircraft's flight path. Abandon the approach unless a landing is ensured well before reaching the intersection.

17. What procedures should be utilized in avoiding wake turbulence when taking off? (AIM 7-3-6)

a. Departing behind a larger aircraft: note the larger aircraft's rotation point, rotate prior to larger aircraft's rotation point. Continue climb above the larger aircraft's climb path until turning clear of its wake.

b. Intersection takeoffs on the same runway: be alert to adjacent larger aircraft operations, particularly upwind of your runway. If intersection takeoff clearance is received, avoid a subsequent heading which will cross below the larger aircraft's path.

c. Departing or landing after a larger aircraft executing a low approach, missed approach or touch-and-go landing: ensure that an interval of at least 2 minutes has elapsed before you take off or land. Because vortices settle and move laterally near the ground, the vortex hazard may continue to exist along the runway, particularly in light quartering wind situations.

d. Enroute VFR (thousand foot altitude plus 500 feet): avoid flight below and behind a large aircraft's path. If a larger aircraft is observed above on the same track (meeting or overtaking), adjust your position laterally, preferably upwind.

18. Who is responsible for wake turbulence avoidance, the pilot or the air traffic controller? (AIM 7-3-8)

The pilot is responsible. Acceptance of instructions from ATC (traffic information, follow an aircraft, visual approach clearance), is acknowledgment that the pilot has accepted responsibility for his/her own wake turbulence separation.

19. Define the term "hydroplaning."

Hydroplaning occurs when the tires are lifted off a runway surface by the combination of aircraft speed and a thin film of water present on the runway.

20. What are the three basic types of hydroplaning?

Dynamic—Occurs when there is standing water on the runway surface. Water about $\frac{1}{10}$-inch deep acts to lift the tire off the runway. The minimum speed at which dynamic hydroplaning occurs has been determined to be about 9 times the square root of the tire pressure in pounds per square inch.

Viscous—Occurs as a result of the viscous properties of water. A very thin film of fluid cannot be penetrated by the tire and the tire consequently rolls on top of the film. Viscous hydroplaning can occur at much slower speeds than dynamic hydroplaning but requires a smooth acting surface.

Reverted Rubber Hydroplaning—Occurs when a pilot, during the landing roll, locks the brakes for an extended period of time while on a wet runway. The friction creates heat which, combined with water, creates a steam layer between the aircraft tire and runway surface.

21. What is the best method of speed reduction if hydroplaning is experienced on landing?

Aerodynamic braking is the most effective means of dealing with a hydroplaning situation. Use of flaps, increased angle of attack, spoilers, reverse thrust, etc., will produce more desirable results than braking.

22. What are several type of illusions in flight which may lead to errors in judgment on landing? (AIM 8-1-5)

Runway width illusion—Narrower than usual runway creates illusion aircraft is higher than actual; pilot tends to fly a lower approach than normal.

Continued

Runway and terrain slope illusion—Upsloping runway/terrain creates illusion aircraft is higher than actual; pilot tends to fly a lower approach than normal. Downsloping runway/terrain has the opposite effect.

Featureless terrain illusion—An absence of ground features creates illusion that aircraft is higher than actual; pilot tends to fly a lower approach than normal.

Atmospheric illusions—Rain on windscreen creates illusion of greater height; atmospheric haze creates illusion of greater distance from runway; pilot tends to fly a lower approach than normal.

23. What is the most effective method of scanning for other air traffic? (AIM 8-1-6)

Effective scanning is accomplished with a series of short, regularly spaced eye movements that bring successive areas of the sky into the central vision field. Each movement should not exceed 10°, and each area should be observed for at least 1 second to enable detection. Although horizontal back and forth eye movements seem preferred by most pilots, each pilot should develop a comfortable scanning pattern and then adhere to it to ensure optimum scanning.

F. High-Altitude Operations

1. What are some basic operational advantages when conducting high-altitude operations?

a. True airspeeds increase with altitude

b. Winds aloft are stronger providing tailwind opportunities

c. Capability to see and avoid thunderstorms

d. Better visibility

e. Less turbulence

f. Above the weather instead of in it

g. Reduced chance for icing

h. Conflicts with other air traffic reduced

2. What are the regulations concerning use of supplemental oxygen on board an aircraft? (14 CFR 91.211)

No person may operate a civil aircraft of U.S. registry:

a. At cabin pressure altitudes above 12,500 feet MSL up to and including 14,000 feet MSL, unless, for that part of the flight at those altitudes that is more than 30 minutes, the required minimum flight crew is provided with and uses supplemental oxygen.

b. At cabin pressure altitudes above 14,000 feet MSL, unless the required flight crew is provided with and uses supplemental oxygen for the entire flight time at those altitudes.

c. At cabin pressure altitudes above 15,000 feet MSL, unless each occupant is provided with supplemental oxygen.

3. What are the regulations pertaining to the use of supplemental oxygen on board a "pressurized" aircraft? (14 CFR 91.211)

Above Flight Level 250:
At least a ten-minute supply of supplemental oxygen, in addition to any oxygen required to satisfy 14 CFR §91.211, is available for each occupant of the aircraft for use in the event that a descent is necessitated by loss of cabin pressurization.

Above Flight Level 350:
At least one pilot at the controls of the airplane is wearing and using an oxygen mask that is secured and sealed that either supplies oxygen at all times or automatically supplies oxygen whenever the cabin pressure altitude of the airplane exceeds 14,000 feet (MSL).

Note: One pilot need not wear and use an oxygen mask while at or below Flight Level 410 if two pilots are at the controls and each pilot has a quick donning type of oxygen mask that can be placed on the face within 5 seconds. Also, if for any reason at any time it is necessary for one pilot to leave the controls of the aircraft when operating at altitudes above Flight Level 350, the remaining pilot at the controls shall put on and use an oxygen mask until the other pilot has returned to that crewmember's station.

4. What are the requirements to operate within Class A airspace? (14 CFR 91.135)

a. Operated under IFR at a specific flight level assigned by ATC;

b. Equipped with instruments and equipment required for IFR operations;

c. Flown by a pilot rated for instrument flight; and

d. Equipped, when in Class A airspace, with:

    i. A radio providing direct pilot/controller communication on the frequency specified by ATC in the area concerned; and

    ii. The applicable equipment specified in 14 CFR §91.215 (transponder regulations).

5. What additional equipment is required when operating above Flight Level 240? (14 CFR 91.205)

For flight at and above 24,000 feet MSL: if VOR navigational equipment is required (appropriate to the ground facilities to be used) no person may operate a U.S.-registered civil aircraft within the 50 states and the District of Columbia at or above FL240 unless that aircraft is equipped with approved distance measuring equipment (DME).

6. What type of navigational charts are utilized when operating at altitudes above 18,000 feet? (AIM 9-1-4)

Enroute high altitude charts are designed for navigation at or above 18,000 feet MSL. This four-color chart series includes the jet-route structure; VHF NAVAIDs with frequency identification, channel, geographic coordinates; selected airports, reporting points. These charts are revised every 56 days.

G. National Transportation Safety Board

1. When is immediate notification to the NTSB required? (NTSB Part 830)

The operator of an aircraft shall immediately, and by the most expeditious means available, notify the nearest NTSB field office when an aircraft accident or any of the following listed incidents occur:

a. Flight control system malfunction

b. Crewmember unable to perform normal duties

c. Turbine engine failure of structural components

d. In-flight fire

e. Aircraft collision in-flight

f. Property damage, other than aircraft, estimated to exceed $25,000

g. Overdue aircraft (believed to be in an accident)

2. After an accident or incident has occurred, how soon must a report be filed with the NTSB? (NTSB Part 830)

The operator shall file a report on NTSB Form 6120.1 or 6120.2, available from NTSB field offices or from the NTSB, Washington D.C., 20594:

a. Within 10 days after an accident;

b. When, after 7 days, an overdue aircraft is still missing.

A report on an "Incident" for which notification is required as described shall be filed only as requested by an authorized representative of the NTSB.

3. Define "aircraft accident." (NTSB Part 830.2)

An aircraft accident means an occurrence associated with the operation of an aircraft which takes place between the time any person boards the aircraft with the intention of flight and all such persons have disembarked, and in which any person suffers death or serious injury, or in which the aircraft receives substantial damage.

4. Define "aircraft incident." (NTSB Part 830.2)

An aircraft incident means an occurrence other than an accident associated with the operation of an aircraft, which affects or could affect the safety of operations.

5. Define the term "serious injury." (NTSB Part 830.2)

Serious injury means any injury which:

a. Requires hospitalization for more than 48 hours, commencing within 7 days from the date the injury was received;

b. Results in a fracture of any bone (except simple fractures of fingers, toes or nose);

c. Causes severe hemorrhages, nerve, muscle or tendon damage;

d. Involves any internal organ; or

e. Involves second- or third-degree burns affecting more than 5% of the body surface.

6. Define the term "substantial damage." (NTSB Part 830.2)

Substantial damage means damage or failure which adversely affects structural strength, performance or flight characteristics of the aircraft and which normally requires major repair or replacement of the affected component.

7. Will notification to the NTSB always be necessary in any aircraft "accident" even if there were no injuries? (NTSB Part 830)

Refer to the definition of "accident." An aircraft accident can be substantial damage and/or injuries, and the NTSB always requires a report if this is the case.

8. Where are accident or incident reports filed? (NTSB Part 830)

The operator of an aircraft shall file any report with the field office of the Board nearest the accident or incident. The National Transportation Safety Board field offices are listed in the U.S. government pages of telephone directories in major cities.

Additional Study Questions

1. **What types of airspace is categorized as "Other" airspace?** (AIM 3-5-1 through 3-5-6)

2. **What wind condition prolongs the hazards of wake turbulence on a landing runway for the longest period of time?** (AIM 7-3-4)

3. **What situation would result in "reverse sensing" of a VOR receiver?** (AC 61-23C)

4. **What is the main difference between controlled airspace and uncontrolled airspace?** (AIM 3-2-1 and 3-3-1)

5. **Do regulations permit a pilot to drop objects from an aircraft in flight?** (14 CFR §91.15)

6. **When a control tower closes, the airspace within the airport surface area reverts from Class D airspace to Class E airspace. How will this affect the VFR weather minimums?** (AIM 3-2-5)

7. **When transitioning busy terminal airspace, where can information be found concerning VFR flyways, corridors and transition routes?** (AIM 3-5-5)

8. **What is the width of a Federal airway?** (14 CFR Part 71)

9. **What significance does an underlined VOR station frequency have on a VFR sectional chart?** (AC 61-23C)

10. **When would a pilot use the following charts: WAC charts; TAC charts; Sectional charts?** (AIM 9-1-4)

Night Flight
Operations

7

A. Night Vision

1. Name the two distinct types of light-sensitive cells located in the retina of the eye. (AC 61-23C)

Rods and cones are the light-sensitive cells located in the retina.

2. What is the function of the cones, and where are they located in the eye? (AC 61-23C)

The cones are used to detect color, detail and far-away objects and are located in the center of the retina at the back of the eye. They are less sensitive to light, require higher levels of intensity to become active, and are most useful in the daylight hours.

3. What is the function of the rods, and where are they located in the eye? (AC 61-23C)

The rods are used for peripheral vision and are located in a ring around the cones. Rods are highly sensitive and are activated by a minimum amount of light. Thus they are better suited for night vision.

4. What is the average time it takes for the rods and cones to become adapted to darkness? (FAA-H-8083-3)

The cones will take approximately 10–15 minutes to adjust to darkness. The rods will take approximately 30 minutes to adjust to darkness.

5. What should the pilot do to accommodate changing light conditions? (FAA-H-8083-3)

The pilot should allow enough time for the eyes to become adapted to the low light levels and then should avoid exposure to bright light which could cause temporary blindness.

6. **When approaching a well-lit runway surrounded by a dark area with little or no features, what illusion should a pilot be alert for?** (AIM 8-1-5)

    Featureless terrain illusion—creates the illusion the aircraft is at a higher altitude than it actually is. The pilot who does not recognize this will fly a lower approach, possibly landing short of the runway.

7. **What can the pilot do to improve the effectiveness of vision at night?** (FAA-H-8083-3)

    a. If possible, allow 30 minutes for the eyes to adjust to the darkness.

    b. Adjust the cockpit lights for optimum night vision and avoid bright levels of light.

    c. Never look directly at an object; try to look off center since this allows better visual perception.

    d. Be aware that depth perception is inhibited due to the lack of visual cues; thus attention to airspeed, altitude, sink rates, and the attitude indicator must be maintained.

    e. Maintain a safe altitude until airport lighting and the airport itself are identifiable and visible. Many pilots have mistaken lighted highways for airports.

    f. Avoid taking medication, drugs and alcohol prior to or during aircraft flights.

B. Pilot Equipment

1. **What equipment should the pilot have for night flight operations?** (FAA-H-8083-3)

    Pilots should carry two flashlights, one white for inspecting the aircraft during preflight and one red during the flight for examining charts. The red light will not affect night vision as it does not glare. However, be aware that items printed in red on aeronautical charts may disappear.

2. What other items should the pilot have on board for night flights? (FAA-H-8083-3)

Pilots should have appropriate navigational charts, including any charts adjacent to the intended route of flight, on board for night flights. Furthermore, these charts should be mounted on a clipboard or mapboard to prevent being lost in a dark cockpit.

C. Aircraft Equipment and Lighting

1. What equipment is required on an aircraft for night flights? (14 CFR 91.205)

In addition to the required equipment for VFR flight during the day as detailed in 14 CFR §91.205(b), the following equipment is required for VFR flight at night:

a. Approved position lights;

b. An approved aviation red or aviation white anticollision light system;

c. If the aircraft is operated for hire, one electric landing light;

d. An adequate source of electrical energy for all installed electrical and radio equipment; and

e. One spare set of fuses, or three spare fuses of each kind required, accessible to the pilot in flight.

2. Explain the arrangement and interpretation of the position lights on an aircraft. (FAA-H-8083-3)

A red light is located on the left wingtip, a green light is located on the right wingtip and a white light is located on the tail. If the pilot observes both a green and red light on another aircraft, then the other aircraft is generally approaching the pilot's position. If the pilot only sees a green light, then the other aircraft is moving left to right in relation to the pilot's position. If the pilot only sees a red light, then the aircraft is moving right to left in relation to the pilot's position.

3. **Position lights are required to be on during what period of time?** (14 CFR 91.209)

   No person may operate an aircraft during the period from sunset to sunrise unless the aircraft has lighted position lights.

4. **When operating an aircraft in, or in close proximity to, a night operations area, what is required of an aircraft?** (14 CFR 91.209)

   The aircraft must

   a. Be clearly illuminated,

   b. Have lighted position lights,

   c. Be in an area which is marked by obstruction lights.

5. **Are aircraft anticollision lights required to be on during night flight operations?** (14 CFR 91.209)

   Yes; however, if the pilot-in-command determines that because of operating conditions it would be in the interest of safety, the anticollision lights can be turned off.

D. Airport and Navigational Aid Lighting

1. **What are the different types of rotating beacons used to identify airports?** (AIM 2-1-8)

   a. White and green—lighted land airport

   b. *Green alone—lighted land airport

   c. White and yellow—lighted water airport

   d. *Yellow alone—lighted water airport

   e. Green, yellow, and white—lighted heliport

   f. White (dual peaked) and green—military airport

   Note: *Green alone or yellow alone is used only in connection with a white and green or white and yellow beacon display respectively.

2. Describe several types of aviation obstruction lighting. (AIM 2-2-3)

a. *Aviation red obstruction lights*—Flashing aviation red beacons and steady burning aviation red lights.

b. *High intensity white obstruction lights*—Flashing high intensity white lights during daytime with reduced intensity for twilight and nighttime operation.

c. *Dual lighting*—A combination of flashing aviation red beacons and steady burning aviation red lights for nighttime operation, and flashing high intensity white lights for daytime operation.

3. What color are runway edge lights? (AIM 2-1-4)

The runway edge lights are white. On instrument runways, however, yellow replaces white on the last 2,000 feet or half the runway length, whichever is less, to form a caution zone for landings.

4. What color are the lights marking the ends of the runway? (AIM 2-1-4)

The lights marking the ends of the runway emit red light toward the runway to indicate the end of the runway to a departing aircraft and emit green outward from the runway end to indicate the threshold to landing aircraft.

5. Describe runway end identifier lights (REIL). (AIM 2-1-3)

REILs are installed at many airports to provide rapid and positive identification of the approach end of a particular runway. The system consists of a pair of synchronized flashing lights located laterally on each side of the runway threshold.

6. What color are taxiway edge lights? (AIM 2-1-9)

Taxiway edge lights are used to outline the edges of taxiways during periods of darkness or restricted visibility conditions. These fixtures emit blue light.

7. What color are taxiway centerline lights? (AIM 2-1-9)

Taxiway centerline lights are steady burning and emit green light.

8. How does a pilot determine the status of a light system at a particular airport? (FAA-H-8083-3)

The pilot needs to check the Airport/Facility Directory and any Notices to Airmen (NOTAMs) to find out about available lighting systems, light intensities and radio-controlled light system frequencies.

9. How does a pilot activate a radio-controlled runway light system while airborne? (AIM 2-1-7)

The pilot activates radio-controlled lights by keying the microphone on a specified frequency. The following sequence can be used for typical radio controlled lighting systems:

a. On initial arrival, key the microphone seven times to turn the lights on and achieve maximum brightness;

b. If the runway lights are already on upon arrival, repeat the above sequence to ensure a full 15 minutes of lighting; then

c. The intensity of the lights can be adjusted by keying the microphone five or three times within 5 seconds.

Additional Study Questions

1. During preflight, what actions should be accomplished to adequately prepare for a night flight? (FAA-H-8083-3)

2. If an engine failure occurs at night, what procedures should be followed? (FAA-H-8083-3)

3. Describe some of the illusions that may occur when flying at night. (AIM 8-1-5)

4. What type of scanning procedure is most effective when flying at night? (FAA-H-8083-3)

5. What type of lighting is used to indicate the runway threshold? runway end? (FAA-H-8083-3)

6. Discuss the meaning of the abbreviations ALSF1, ALSF2, MALSR, etc., found in the runway data section of an Airport/Facility Directory? (AIM 2-1-1)

7. What precautionary actions can you take, at night, on the ground, to warn others your aircraft's propeller is or will be in operation? (FAA-H-8083-3)

8. Discuss the FAA safety program: "Operations Lights On." (FAA-H-8083-3)

9. While flying at night you observe directly ahead of your aircraft and at the same altitude, a steady red light directly to the right of a steady green light. What action should you take? (FAA-H-8083-3)

10. Where can a pilot find information on the availability and status of airport lighting?

Aeromedical
Factors

8

A. Fitness for Flight

1. What regulations apply to medical certification?

Part 67—Medical Standards and Certification.

2. As a flight crewmember, you discover you have high blood pressure. You are in possession of a current medical certificate. Can you continue to exercise the privileges of your certificate? (AIM 8-1-1)

No, the regulations prohibit a pilot who possesses a current medical certificate from performing crewmember duties while the pilot has a known medical condition or an increase of a known medical condition that would make the pilot unable to meet the standards for the medical certificate.

3. Are flight crewmembers allowed the use of any medications while performing required duties? (AIM 8-1-1)

The regulations prohibit pilots from performing crewmember duties while using any medication that affects the faculties in any way contrary to safety. The safest rule is not to fly as a crewmember while taking any medication, unless approved to do so by the FAA.

4. Are there any over-the-counter medications that could be considered safe to use while flying? (AIM 8-1-1)

No; pilot performance can be seriously degraded by both pre-scribed and over-the-counter medications, as well as by the medical conditions for which they are taken. Many medications have primary effects that may impair judgment, memory, alertness, coordination, vision, and the ability to make calculations. Also, any medication that depresses the central nervous system can make a pilot more susceptible to hypoxia.

5. **What are several factors which may contribute to impairment of a pilot's performance?** (AIM 8-1-1)

I llness
M edication
S tress
A lcohol
F atigue
E motion

B. Flight Physiology

1. What is hypoxia? (AIM 8-1-2)

Hypoxia is a state of oxygen deficiency in the body sufficient to impair functions of the brain and other organs.

2. Where does hypoxia usually occur, and what symptoms should one expect? (AIM 8-1-2)

Although a deterioration in night vision occurs at a cabin pressure altitude as low as 5,000 feet, other significant effects of altitude hypoxia usually do not occur in the normal healthy pilot below 12,000 feet. From 12,000 feet to 15,000 feet of altitude, judgment, memory, alertness, coordination, and ability to make calculations are impaired, and headache, drowsiness, dizziness and either a sense of well being or belligerence occur.

3. What factors can make a pilot more susceptible to hypoxia? (AIM 8-1-2)

a. Carbon monoxide inhaled in smoking or from exhaust fumes.
b. Anemia (lowered hemoglobin)
c. Certain medications
d. Small amounts of alcohol
e. Low doses of certain drugs (antihistamines, tranquilizers, sedatives, analgesics, etc.)

Also, extreme heat or cold, fever, and anxiety increase the body's demand for oxygen, and hence its susceptibility to hypoxia.

4. How can hypoxia be avoided? (AIM 8-1-2)

Hypoxia is prevented by heeding factors that reduce tolerance to altitude; by enriching the inspired air with oxygen from an appropriate oxygen system; and by maintaining a comfortable, safe cabin pressure altitude. For optimum protection, pilots are encouraged to use supplemental oxygen above 10,000 feet during the day and above 5,000 feet at night.

5. What is hyperventilation? (AIM 8-1-3)

Hyperventilation, or an abnormal increase in the volume of air breathed in and out of the lungs, can occur subconsciously when a stressful situation is encountered in flight. This results in a significant decrease in the carbon dioxide content of the blood. Carbon dioxide is needed to automatically regulate the breathing process.

6. What symptoms can a pilot expect from hyperventilation? (AIM 8-1-3)

As hyperventilation "blows off" excessive carbon dioxide from the body, a pilot can experience symptoms of lightheadedness, suffocation, drowsiness, tingling in the extremities, and coolness, and react to them with even greater hyperventilation. Incapacitation can eventually result from uncoordination, disorientation, and painful muscle spasms. Finally, unconsciousness can occur.

7. How can a hyperventilating condition be reversed? (AIM 8-1-3)

The symptoms of hyperventilation subside within a few minutes after the rate and depth of breathing are consciously brought back under control. The buildup of carbon dioxide in the body can be hastened by controlled breathing in and out of a paper bag held over the nose and mouth.

8. What is "ear block"? (AIM 8-1-2)

As the aircraft cabin pressure decreases during ascent, the expanding air in the middle ear pushes open the Eustachian tube and escapes down to the nasal passages, thereby equalizing in pressure with the cabin pressure. But this isn't automatic during descent, and the pilot must periodically open the Eustachian tube to equalize pressure. An upper respiratory infection or a nasal allergic condition can produce enough congestion around the Eustachian tube to make equalization difficult. Consequently, the difference in pressure between the middle ear and aircraft cabin can build to a level that holds the Eustachian tube closed, making equalization difficult if not impossible. An ear block produces severe pain and loss of hearing that can last from several hours to several days.

9. How is ear block normally prevented from occurring? (AIM 8-1-2)

Normally this can be accomplished by swallowing, yawning, tensing muscles in the throat or, if these do not work, by the combination of closing the mouth, pinching the nose closed and attempting to blow through the nostrils (Valsalva maneuver). It is also prevented by not flying with an upper respiratory infection or nasal allergic condition.

10. What is "spatial disorientation"? (AC 61-27C)

The inability to determine accurately the attitude or motion of the aircraft in relation to the Earth's surface.

11. What causes spatial disorientation? (AC 61-27C)

Various complex motions, forces, and complex visual scenes encountered in flight under adverse weather conditions or at night can create illusions of motion and position.

12. What is the cause of motion sickness, and what are its symptoms? (FAA-H-8083-3)

Motion sickness is caused by continued stimulation of the tiny portion of the inner ear which controls the pilot's sense of balance. The symptoms are progressive. First, the desire for food is lost. Then saliva collects in the mouth and the person begins to perspire freely. Eventually, he or she becomes nauseated and disoriented. The head aches and there may be a tendency to vomit. If the air sickness becomes severe enough, the pilot may become completely incapacitated.

13. What action should be taken if a pilot or his passenger suffers from motion sickness? (AC 61-23C)

If suffering from airsickness while piloting an aircraft, open up the air vents, loosen the clothing, use supplemental oxygen, and keep the eyes on a point outside the airplane. Avoid unnecessary head movements. Then cancel the flight and land as soon as possible.

14. What regulations apply, and what common sense should prevail concerning the use of alcohol? (AIM 8-1-1)

The regulations prohibit pilots from performing crewmember duties within 8 hours after drinking any alcoholic beverage or while under the influence of alcohol. However, due to the slow destruction of alcohol, a pilot may still be under influence 8 hours after drinking a moderate amount of alcohol. Therefore, an excellent rule is to allow at least 12 to 24 hours from "bottle to throttle," depending on the amount of alcoholic beverage consumed.

15. What is carbon monoxide poisoning? (FAA-H-8083-3)

Carbon monoxide is a colorless, odorless and tasteless gas contained in exhaust fumes. When breathed, even in minute quantities over a period of time, it can significantly reduce the ability of the blood to carry oxygen. Consequently, effects of hypoxia occur.

16. How does carbon monoxide poisoning occur, and what symptoms should a pilot be alert for? (AIM 8-1-4)

Most heaters in light aircraft work by air flowing over the manifold. Use of these heaters while exhaust fumes are escaping through manifold cracks and seals is responsible for several nonfatal and fatal aircraft accidents from carbon monoxide poisoning each year. A pilot who detects the odor of exhaust or experiences symptoms of headache, drowsiness, or dizziness while using the heater should suspect carbon monoxide poisoning.

17. What action should be taken if a pilot suspects carbon monoxide poisoning? (AIM 8-1-4)

A pilot who suspects this condition to exist should immediately shut off the heater and open all air vents. If symptoms are severe, or continue after landing, medical treatment should be sought.

18. What precautions should be taken before flight if you or your passengers have been involved in recent scuba diving activity? (AIM 8-1-2)

A pilot or passenger who intends to fly after scuba diving should allow the body sufficient time to rid itself of excess nitrogen absorbed during diving. If not, decompression sickness due to evolved gas can occur during exposure to higher altitudes and create a serious in-flight emergency.

The recommended waiting time before flight to cabin pressure altitudes of 8,000 feet or less is at least 12 hours after diving which has not required controlled ascent (non-decompression diving) and at least 24 hours after diving which has required controlled ascent (decompression diving). The waiting time before flight to cabin pressure altitudes above 8,000 feet should be at least 24 hours after any scuba diving.

C. Aeronautical Decision Making

1. What method does the FAA encourage pilots to use as a logical way to approach decision making? (AC 60-22)

The DECIDE Model is a six-step, continuous-loop decision-making process which can be used to assist a pilot when he/she is faced with a situation requiring judgment:

a. **Detect**—The decisionmaker detects the fact that change has occurred.

b. **Estimate**—The decisionmaker estimates the need to counter or react to the change.

c. **Choose**—The decisionmaker chooses a desirable outcome (in terms of success for the flight).

d. **Identify**—The decisionmaker identifies actions that could successfully control the change.

e. **Do**—The decisionmaker takes the necessary action.

f. **Evaluate**—The decisionmaker evaluates the effect(s) of his/her action countering the change.

2. What are the 5 types of hazardous attitudes the FAA has identified and provided antidotes for, to encourage pilots to develop a realistic perspective on attitudes toward flying? (AC 60-22)

Antiauthority (Don't tell me!)—Follow the rules, they are usually right.

Impulsivity (Do something quickly!)—Not so fast. Think first.

Invulnerability (It won't happen to me.)—It could happen to me.

Macho (I can do it.)—Taking chances is foolish.

Resignation (What's the use?)—I'm not helpless, I can make a difference.

Additional Study Questions

1. What is acute fatigue? Chronic fatigue? (AIM 8-1-1)

2. What is stress? (AIM 8-1-1)

3. What are some examples of the different types of spatial disorientation? (AIM 8-1-5)

4. How can spatial disorientation be prevented? (AIM 8-1-5)

5. What is the difference between anemic hypoxia and hypoxic hypoxia? (AIM 8-1-2)

6. What two common "stimulant" type drugs are used widely but not prohibited by regulations? (AC 67-2)

7. How can a pilot detect the presence of carbon monoxide in the cockpit? (AC 20-32)

8. What is "sinus block"? (AIM 8-1-2)

9. Rain on the windshield of an aircraft on approach to landing will produce what type of illusion? (AIM 8-1-6)

10. What actions can you take to reduce the chances that a passenger will suffer from motion sickness? (AC 67-2)

Commercial
Flight
Maneuvers

A. Steep Turns

1. What is a "steep turn"? (FAA-H-8083-3)

Steep turns are those resulting from a degree of bank (more than approximately 45 degrees) at which the overbanking tendency of an airplane overcomes stability, and the bank tends to increase unless pressure is applied to the aileron controls to prevent it. Maximum turning performance is attained and relatively high load factors are imposed.

2. What is the desired bank angle in a steep turn? (FAA-S-8081-12A)

A bank angle of 50° (±5°) is desired.

3. What is the recommended entry speed for a steep turn? (FAA-S-8081-12A)

Establish the manufacturer's recommended entry speed or a speed which will not exceed the design maneuvering speed, plus or minus 5 knots.

4. How do you maintain altitude in a steep turn? (FAA-H-8083-3)

To maintain altitude as well as orientation requires an awareness of the relative position of the nose, the horizon, the wings, and the amount of turn. If the altitude begins to increase, the bank should be increased by coordinated use of aileron and rudder. If the altitude begins to decrease, the bank should be decreased by coordinated use of aileron and rudder.

B. Chandelles

1. What is a "chandelle"? (FAA-H-8083-3)

A "chandelle" is a maximum performance climbing turn beginning from approximately straight and level flight, and ending at the completion of 180 degrees of turn in a wings-level, nose-high attitude at the minimum controllable airspeed.

2. In a chandelle, constant bank and changing pitch occur in what part of the maneuver? (FAA-S-8081-12A)

The first 90° of turn require a constant 30° of bank and a gradual and constant change in pitch attitude.

3. In a chandelle, constant pitch and changing bank occur in what part of the maneuver? (FAA-S-8081-12A)

The last 90° of turn requires a very gradual change in bank from 30° to 0° and a constant pitch attitude so as to arrive at minimum airspeed as the airplane is rolled out to a wings-level attitude.

4. What is the maximum amount of bank in the chandelle? (FAA-S-8081-12A)

30° of bank.

5. What should your speed be upon completion of the chandelle? (FAA-S-8081-12A)

You should begin a coordinated constant rate rollout from the 90-degree point to the 180-degree point maintaining specified power and a constant pitch attitude that will result in approximately 1.2 V_{S1}, plus or minus 5 knots.

6. At what two points will your wings be level in a chandelle? (FAA-S-8081-12A)

Immediately before entering the chandelle and upon rollout at the 180-degree point.

C. Lazy Eights

1. What is a "lazy eight"? (FAA-H-8083-3)

A lazy eight consists of two 180-degree turns, in opposite directions, while making a climb and a descent in a symmetrical pattern during each of the turns.

2. Where should the wind be when beginning a lazy eight? (FAA-H-8083-3)

The maneuver should be entered into the wind to avoid drifting too far from the area originally cleared for the maneuver.

3. Where should the highest pitch attitude occur in a lazy eight? (FAA-H-8083-3)

At the 45-degree point the pitch attitude should be at a maximum and the angle of bank continuing to increase. Also, at the 45-degree point, the pitch attitude should start to decrease slowly toward the horizon at the 90-degree reference point.

4. Where should the lowest nose-down attitude occur in a lazy eight? (FAA-H-8083-3)

When the airplane has turned 135°, the nose should be at its lowest pitch attitude.

5. What are the altitude, airspeed and heading tolerances allowed when performing a lazy eight? (FAA-S-8081-12A)

You should begin a coordinated constant rate of rollout from the 90-degree point to the 180-degree point maintaining specified power and pitch attitude that will result in a rollout within ±10 degrees of the desired heading and airspeed within ±5 knots of power-on stall speed.

6. To summarize, describe the appropriate values to be obtained in a lazy eight at the entry, 45-, 90-, 135- and 180-degree points. (FAA-H-8083-3)

Entry
—Level flight
—Maneuvering or cruise speed (whichever is less) or
    manufacturer's recommended speed

45-degree point
—Maximum pitch-up attitude
—Bank angle at 15°

Continued

90-degree point
—Bank angle approximately 30°
—Minimum airspeed
—Maximum altitude
—Level pitch attitude

135-degree point
—Maximum pitch-down attitude
—Bank approximately 15°

180-degree point
—Level flight
—Original heading (±10°)
—Entry airspeed (±10 knots)
—Entry altitude (±100 feet)

D. Eights-On-Pylons

1. What are "eights-on-pylons"? (FAA-H-8083-3)

Eights-on-pylons is a training maneuver which involves flying the airplane in circular paths, alternately left and right, in the form of a figure 8 around two selected points or pylons on the ground. No attempt is made to maintain a uniform distance from the pylon. Instead, the airplane is flown at such an altitude and airspeed that a line parallel to the airplane's lateral axis, and extending from the pilot's eye appears to pivot on each of the pylons.

2. How do you determine pivotal altitude for eights-on-pylons? (FAA-H-8083-3)

An entry pivotal altitude value is initially calculated by using the ground speed entering the maneuver. Since the first turn is made into the wind, this will be the slowest ground speed. Use the formula:

If airspeed indicator is in MPH:
   $GS^2 \div 15 = PA$

If airspeed indicator is in KNOTS:
   $GS^2 \div 11.3 = PA$

3. Does the pivotal altitude change in eights-on-pylons? (FAA-H-8083-3)

Yes, the pivotal altitude is critical and will change with variations in ground speed. Since the headings throughout the turns continually vary from directly downwind to directly upwind, the ground speed will constantly change. This will result in the proper pivotal altitude varying slightly throughout the eight. Therefore, adjustment must be made for this by climbing and descending as necessary to hold the reference line or point on the pylons. This change in altitude will be dependent on how much the wind affects the ground speed.

Remember...
Ground speed goes UP, Pivotal Altitude goes UP.

Ground speed goes DOWN, Pivotal Altitude goes DOWN.

4. How far should one pylon be from the other pylon in eights-on-pylons? (FAA-S-8081-12A)

The pylons should be of sufficient distance apart to permit approximately 3 to 5 seconds of straight and level flight between them.

5. Where is the highest pivotal altitude likely to occur in eights-on-pylons? (FAA-H-8083-3)

As the airplane turns downwind the ground speed increases; consequently the pivotal altitude is higher and the airplane must climb to hold the reference line on the pylon.

6. Where is the lowest pivotal altitude likely to occur in eights-on-pylons? (FAA-H-8083-3)

As the airplane heads into the wind, the ground speed decreases; consequently the pivotal altitude is lower and the airplane must descend to hold the reference line on the pylon.

7. **What action should you take if your wing reference point appears to move ahead of the pylon? Move behind the pylon?** (FAA-H-8083-3)

   If the reference line appears to move ahead of the pylon, the pilot should increase altitude. If the reference line appears to move behind the pylon, the pilot should decrease altitude. Varying rudder pressure to yaw the airplane and force the wing and reference line forward or backward to the pylon is a dangerous technique and must not be attempted.

Additional Study Questions

1. **Why does the airplane have a tendency to overbank in a steep turn?** (FAA-H-8083-3)

2. **What effect will a steep turn have on an airplane's stall speed and load factor? Why?** (FAA-H-8083-3)

3. **Where should the wind be when beginning a chandelle? Why?** (FAA-H-8083-3)

4. **What altitude and airspeed will you use when entering a chandelle?** (FAA-H-8083-3)

5. **When entering a lazy eight, will you establish the bank first, pitch first, or both at the same time?** (FAA-H-8083-3)

6. **Discuss reference point selection for the lazy eight maneuver.** (FAA-H-8083-3)

7. **When performing a lazy eight, why is more right rudder required in the right turn than in the left turn?** (FAA-H-8083-3)

8. **When performing an eights-on-pylon maneuver, where will the steepest angle of bank occur?** (FAA-H-8083-3)

9. **Will the pivotal altitude change when changing the angle-of-bank in an eights-on-pylon maneuver?** (FAA-H-8083-3)

10. **What initial altitude and airspeed will you use when entering an eights-on-pylon maneuver?** (FAA-H-8083-3)

Notes

Notes